Decision Making and Modelling in Cognitive Science

Sisir Roy

Decision Making and Modelling in Cognitive Science

 Springer

Sisir Roy
National Institute of Advanced Studies, IISc
   Campus
Bengaluru, Karnataka
India

ISBN 978-81-322-3620-7   ISBN 978-81-322-3622-1   (eBook)
DOI 10.1007/978-81-322-3622-1

Library of Congress Control Number: 2016948803

Printed on acid-free paper

This Springer imprint is published by Springer Nature
The registered company is Springer (India) Pvt. Ltd.
The registered company address is: 7th Floor, Vijaya Building, 17 Barakhamba Road, New Delhi 110 001, India

Foreword

Decision making is a hot topic in the cognitive neurosciences, with several applications in the practical human sciences. If humans is not the rational agent assumed by the classics of the science of economics, is he then an irrational one, as some authors have recently claimed?

The answer could be stated in this manner: it may be the case that we are not irrational, but follow a quantum-probabilistic kind of rule. In this book, Sisir Roy deeply investigates the logic of human decision making, arguing that the probability calculus, which is similar to the formalism of quantum theory, better describes and simulates human decision making. Classical logic and probability theory would not account for how human beings really make decisions in everyday life.

If human rationality is to be conceived according to quantum theory, which interpretation of the theory should be selected? There are so many interpretations that, depending on a particular one, this may not be a good choice. The book overcomes this problem by going beyond a purely epistemological view—as in the case of the Copenhagen interpretations–towards an encompassing theory of reality based on quantum theory, i.e., "quantum ontology".

The theoretical path followed by Sisir Roy contains an excursion through quantum probability theory, non-Boolean logic, context dependence, and possible relationships between modern neuroscience and quantum logic. Besides his expertise in formal sciences, he has also collaborated with reputed neuroscientists such as Rudolf Llinás and Gustav Bernroider. This background affords concreteness to the arguments developed in the book, all of them well grounded in contemporary scientific achievements.

As long as the ontology expresses itself in the minds of human beings, there may be cultural patterns that fit the corresponding rules. This seems to be the case for Buddhism, a way of thinking and living that is discussed in the final chapters.

This book is a formidable interdisciplinary investigation into the logic of human behavior that advances our understanding of decision-making processes. It will satisfy the most rigorous specialists while offering a wealth of information to the general reader.

Alfredo Pereira, Jr.
Adjunct Professor
São Paulo State University (UNESP)
São Paulo, Brazil

Preface

> *Hence, in order to have anything like a complete theory of human rationality, we have to understand what role emotion plays in it.*
>
> —Herbert Simon (*Reason in Human Affair*, 1983)

One of the fundamental requirements in the cognitive processes of human beings is to decide with precision. Therefore, it is necessary to understand how human decision makers, in actual situations, i.e., in complex real-world settings, make decisions as well as to learn how to these processes are supported. Though, taking it as a well-established fact, we can describe the main themes of naturalistic decision making as classical decision theory, we should keep in mind some of limitations of that theory. It has been already been recognized that an axiomatic, as well as other kinds of rigorous models of the cognitive decision making, are very much in need. The recent empirical findings in cognitive domain clearly suggest the necessity of changing the paradigm from classical Bayesian probability theory to quantum probability to construct the model of decision making in a consistent manner. Some of these empirical findings are based on gambling. For many centuries across various cultures, gambling has been treated as a form of entertainment. The studies on decision making during gambling, raise much interest regarding the role of interaction between cognition and emotion. The necessity of a theoretical model and its computational aspect are very much thought to answer the question *"how do emotions affect the cognitive function of a decision maker?"* One of the challenging aspects of artificial intelligence (AI) is to model the "human characteristics" like emotions, behavior, etc. in a comprehensive manner. Some attempts have been made to build up theoretical models of emotions in decision making and judgments using multidimensional logic. In this book, the author emphasizes the use of quantum probability, i.e., an extension of quantum logic, to model the decision-making process in the cognitive domain. Now, quantum logic can be shown to be a kind of paraconsistent logic that incorporates the contradictions arising from the simultaneous existence of two mutually exclusive events in a logical way rather than discarding them. This gives rise to a new possibility to model the various degrees of contradictions involved in emotions, as well as to quantify the effect of emotions on judgments and decision making.

The book is planned according to the following scheme: In Chap. 1, various aspects of decision making in the cognitive domain are critically discussed. The role of emotions and logic play very important roles in decision making. They are discussed in the latter half of this chapter.

Various approaches to decision making are discussed in Chap. 2. Two main categories of decision theories, i.e., descriptive and normative theories, are elaborated here. The axiomatic approach deals with deterministic axioms that cannot comply with the uncertainty involved during the decision-making process. In such a situation, the Bayesian framework provides readily applicable statistical procedures where typical inference questions are addressed. Here, the Bayesian probabilistic approach is more appropriate to handle empirical data. Then the importance of the Dumpster-Shafer theory (basically, the extended framework of Bayesian probability theory) is also discussed in handling empirical data. However, this approach is not yet fully developed.

In Chap. 3, decision making and functioning of the brain are discussed from a neuroscience perspective. One of the most challenging aspects of understanding the brain is to understand its predictability. The brain needs to tackle the uncertain situation associated with neuronal dynamics for any kind of decision making. This uncertainty is due to the existence of various types of noise or unwanted variations associated with neuronal functioning. To handle such uncertainties, the Bayesian approach is discussed.

New empirical findings for decision making in the cognitive domain are classified into different categories. We critically analyze this evidence in Chap. 4. The data clearly indicate that classical probability cannot explain the results consistently. Many authors suggested that the concept of quantum probability is needed to explain the data. Of course, this framework of quantum probability is an abstract framework devoid of any material content. Here, the quantum formalism as such is not considered as applicable to the neurophysiology of the brain or in the cognitive domain.

To understand quantum probability, it is required to have some mathematical knowledge of vector spaces, scalar products, operators, Hilbert spaces, etc. These mathematical concepts are described in Chap. 5.

Niels Bohr introduced the concept of the complementary principle in understanding the mutually exclusive aspects of an entity in microscopic domains. For example, the particle and wave aspects of a microscopic entity like the electron are two mutually exclusive aspects, i.e., they cannot be measured simultaneously with infinite precision. This concept of complementarity is very important in the context of the total probability sum rule. Along with a collaborator, I have proposed a generalized complementary principle in the cognitive domain. This is discussed in Chap. 6.

Quantum probability is an extension of quantum logic. This is different from Boolean logic. In Boolean logic, there exist two truth values, 'yes or no' (1 or 0).

On the other hand, in quantum mechanics, the intermediate situation due to the superposition rule poses contradictions within the purview of Boolean logic. There is a long debate over whether one can think of logic as separate from cognition or psychology. In the Bayesian framework, probability is considered to be an extension of logic. The classical Bayesian probability is used to handle the uncertain situation for decision making in the cognitive domain. Here, since the concept of quantum probability is used to explain the data, we discuss the structure of quantum logic in Chap. 7.

Quantum ontology has been discussed by many authors, since the very inception of quantum theory. Recently, quantum ontology and quantum probability have been defined in an abstract manner, i.e., defined in such a manner that they can be applied to any branch of knowledge like social sciences, biology, etc. But the main challenge is how to contextualize these. For example, to apply them to cognition, one needs to contextualize them in the context of neuroscience. Quantum ontology and its contextualization are discussed in Chap. 8.

In Chap. 9, we discuss quantum logic in the context of modern neuroscience.

Finally, we make some remarks regarding emotions, affective computing, and quantum logic. Quantum logic and decision making raise important epistemological issues, and similar questions are found in ancient Indian texts. These are discussed in Chap. 10.

The framework of quantum probability and quantum logic help us make a detailed analysis of mental functions and their modeling. This will open up new vistas to understand the man-machine interface and affective computing from a more realistic perspective.

The writing of this book would have been impossible without help from my wife, Dr. Malabika Roy, who read the entire manuscript and provided critical suggestions. Some portions of the book were written during my stay at the Indian Statistical Institute, Kolkata, and the rest at the National Institute of Advanced Studies, Bengaluru. The representatives of Springer are greatly acknowledged for their very constructive suggestions from time to time, which ultimately produced the present, final form of this manuscript.

Bengaluru, India   Sisir Roy
2016

Contents

About the Author

Sisir Roy is T.V.Raman Pai Chair Visiting Professor at the National Institute of Advanced Studies, Indian Institute of Science Campus, Bengaluru, India. He was previously professor at the Physics and Applied Mathematics Unit, Indian Statistical Institute, Kolkata, and was associated with that institute for over 30 years. He is a quantum physicist, and his interests include the foundations of quantum theory, cosmology, brain-function modelling, and higher-order cognitive activities. He has published more than 150 papers in peer-reviewed international journals and 12 research monographs/edited volumes with Kluwer Academic Publishers, World Scientific, etc. He has been a visiting professor at George Mason University and the University of Arkansas, and at the Henri Poincare Institute in Paris. His present focus of research is to pursue interdisciplinary research on quantum probability and cognitive science, information theory in living organisms, and neuroscience and consciousness. His forthcoming monographs are on quantum effects in biology and the role of noise in living organisms.

Chapter 1
Introduction

Abstract The purpose of the decision-making process is to determine the results while committing a categorical statement or proposition. A statement or proposition is a sentence which is either true or false. A categorical proposition or statement relates two classes or categories. This process is used widely in many disciplines, for example, in complex scientific, engineering, economical, and management situations. It is necessary to consider all possible rational, heuristic, and intuitive selections so that we can summarize the results in arriving at a decision. The diversified and broad range of interests for understanding this process have induced scientific researcher also to employ a diverse and broad range of research methodologies. They began by exploring other related but independent avenues of thinking, for example, taking into account the many methods of empirical observations, together with developing essentially-related mathematical analysis, including many kinds of computational modelling. Following this mode of search, it would be possible, theoretically, to identify a method for making crucial observations. In turn, its consequences continue to enrich philosophical discourses and to further fragment decision research. Many major attempts have been made to develop independent perspectives connected to various frameworks, such as; game theory, Bayesian models, and expected utility models; models connected to behavioral decision; and approaches related to information processing for neural networks and cognitive architectures. It has already been recognized that axiomatic, as well as other kind of rigorous models of the cognitive decision making, are very much needed. The recent empirical findings in the cognitive domain clearly suggest the necessity of changing the paradigm from classical Bayesian probability theory to quantum probability to construct the model of decision making in a consistent manner. However, quantum probability is an extension of quantum logic which only incorporates the contradictions arising out of the simultaneous existence of two mutually exclusive events in a logical way rather than discarding them. It gives rise to a new possibility to model certain degrees of contradictions involved in emotions, as well as to quantify the effect of emotions on judgments and decision making.

© Springer India 2016
S. Roy, *Decision Making and Modelling in Cognitive Science*,
DOI 10.1007/978-81-322-3622-1_1

Keywords Bayesian probability · Quantum probability · Empirical evidence ·
Normative models · Human cognition · Axiomatic approach · Fisher information ·
Emotions · Affective computation · Quantum logic

1.1 Various Aspects of Decision Making

The process of making a decision is a deliberative process whose ultimate goal is to
determine the results when committing a categorical proposition. The closest
possible analogy to this process can be cited as the considerations made by a judge
or jury who must take time to weigh the evidence in a case. In such a situation, it is
must to gain knowledge about each of the alternative scenario. Not only 'must that',
but all the possible interpretations and/or ramifications are to be settled before
delivering a verdict. The adaptive behavior in humans and other animals is fundamentally determined by the ability to respond and to be its flexible. These special
aspects of the decision process lead to its wide use in the arena, for example, in
complex scientific, engineering, economical, and management situations. Here, it is
necessary to consider all possible rational, heuristic, and intuitive selections so that
we can summarize these results in arriving at a decision. Thus, this process can be
viewed as a tool applied in almost every procedure of daily life. The decision
making, being a basic mental process, surfaces in the course of thinking in human
mind, almost every few seconds, consciously or subconsciously. One of the fundamental requirements in the cognitive processes of human beings is to decide with
precision. It is a deliberative process leading to the resolution of the propositions
that are primarily categorically based on certain criteria. This is true also for artificial systems like autonomous software agents and robots. In other words, this
means that it is always possible to reach a decision among a set of alternatives. For
example, in describing naturalistic decision making, it is necessary to understand
how human decision makers, in actual situations, i.e., in complex real-world settings, make decisions and then to learn how these processes are supported. Though,
taking it as a well-established fact, we can describe the main themes of naturalistic
decision making as a classical decision theory, we should keep in mind some of the
limitations of the theory. Many examples exist that demonstrate the quite diverse
kinds of decisions taken that need explanation.

Many kinds and a wide range of interests have induced scientific researchers to
adopt, likewise, a diverse range of research methodologies. They started exploring
other related, but independent, avenues of thinking, for example, taking into
account many techniques of empirical observations, together with developing
essentially-related mathematical analysis. At the same time, they also adopted
diverse methods of computational modelling to analyze the data obtained from the
experimental observations. This evidently led them to the need for their further
search further for more demanding and critical observations. In turn, this consequently continues to enrich philosophical discourses and the fragmentation of
decision research. Many major attempts have also been made to develop the

potential of some other independent perspectives. For a frame of reference, we mention others of a different nature in this regard, such as: normative (e.g., game theory, Bayesian models, and expected utility models); models connected to behavioral decision (e.g., heuristics, biases, and prospect theory); and approaches related to information processing (e.g., neural networks and cognitive architectures).

Thus, the decision-making process takes into account, very carefully, all the possible evidence and their weights simultaneously. It involves the generation of commitment to a categorical proposition, intended to achieve particular goals. A quite apt analogy can be drawn with the situation where a judge or jury must give sufficient time and consideration to weighing evidence supporting alternative interpretations. Not only that, he/she must take into account of, and/or the possible ramifications, before settling on a verdict. Before taking a decision, a deliberative process mandatory, to be followed by the results in the commitment so that this leads to a categorical proposition, constructed through these processes, before settling on a verdict. Just like that, perceptual tasks start only after the propositions have been made. These facts become quite important in distinguishing between various pieces of sensory evidence. This evidence transiently encodes all the possible information from the senses, including also a decision variable. The role of this variable is to accumulate and store evidence over time, until the final commitment is reached. It is a basic process which, consciously or subconsciously, occurs every few seconds, in the course of thinking in human mind. All these above factors confirm that repetitive application of the fundamental cognitive process is essential for the perception of real-world decisions. These results also show that all categories of decision strategies can be well fitted into the formally described decision processes.

Special aspects of decision making are emphasized in each discipline. For example, the framework of decision-making theory needs to be contextualized in the various disciplines, like statistics, cognitive science, informatics, economics, social psychology, computer science, and artificial intelligence (AI) (Wang and Ruhe 2007; Wang 2003, 2007), as well as in many other practical challenges related to business, politics, investments and insurance, law, medicine, etc., while considering the specific aspects of each discipline. Thus, this wide range of applications has led to a large, diverse set of research methodologies in the field of decision making and judgment (Kahneman 2003). In their book, Bell et al. (1995) have explained in detail the possible kinds of applications of decision making, especially concerning descriptive, normative, and prescriptive interactions–where each of the disciplines emphasize a certain aspect of decision making.

Many important attempts have also been made in the domain of independent perspectives such as the normative approach, Bayesian and utility models (Horvitz 1999; Manski 1977; Manski and Lerman 1977), game theory (Kadane and Larkey 1982), decision theory in behavioral models (heuristics, biases, and prospect theory), and in the framework of information processing (neural networks and the cognitive domain). For example, it is necessary to construct an axiomatic approach to understand decision making in the cognitive domain. An axiomatic approach is a way of describing the probability of even of some axioms or rules, intended to assign probabilities. It has been already been recognized that an axiomatic and rigorous model of the cognitive decision-making needed to arrive at real-world decisions, which are fundamentally the decisions taken in cognitive domain. Within the axiomatic approach (Edwards 1954; Giang and Shenoy 2005), decision theories are often formulated with deterministic axioms. The stochastic variation in the data should not be considered within this framework of deterministic axioms.

Probability theory is employed to understand decision making in a more general way. To start with, a probabilistic model includes the formal characterization of an inductive problem as a start and then specifies the hypotheses under consideration followed by finding the relationship between these hypotheses and the observed data. This leads to the prior probability of each hypothesis. In this manner, a transparent account of the assumptions is considered which enables the problem to be solved and makes it easier to explore the consequences of various assumptions. Basic hypotheses can take any form, for example, the weights in a neural network of structured symbolic representations, as long as they specify a probability distribution over observed data. Likewise, different prior distributions over hypotheses can be assumed by capturing different inductive biases. This approach makes no a priori commitment to any class of representations or inductive biases. On the other hand, it provides a framework for evaluating the various different proposals.

The problem of human cognition is still a puzzling issue to research and to model adequately. Human cognition depends on how the brain behaves in a particular instance, identifies the concerned behavior, and ultimately responds to a signal. Strangely, all these activities are performed among myriads of noise present in the surroundings (called external noise), as well as in the neurons themselves (called internal noise) (Roy and Llinás 2012). Thus it is not surprising to assume that the functionality encompasses various uncertainties. Possibly, a complicated pathway consisting of different types of uncertainties are present in the continuum, which can play a major role in human cognition. Now, whenever considered within the context of living organisms, the term "noise" usually refers to the variance which is always present among measurements. The origin of this noise can be due to the repeated identical experimental conditions or output signals from these systems. It is worth mentioning that both these conditions are universally characterized by the presence of background fluctuations. Previously, noise was generally regarded as a problem, especially in many non-biological systems such as electronics or communication sciences, since primary importance is given to send error-free messages in these application. However, these views have totally changed due to recent developments in diverse technological and biological fields of research. The positive contributive roles of noise has been observed during the development of nanotechnology, information technology, and human genomics. For example, the discovery of stochastic resonances (SR) in non-linear dynamics brought a shift in the perception of noise. Instead of being considered to be a problem, it becomes fundamental to the system function, especially so in biology. But the next question arises: to what extent is the biological system dependent on random noise? As a matter of fact, due to recent developments in neurophysics, it seems feasible into propose that noise also can play an important role, especially when neuronal communication and oscillatory synchronization are concerned. Recently observed and eliciting much attention, many biological systems possess a fractal nature. For example, not only self-similarity in the spatial structure has been observed, but it also has been found that there exists temporal fluctuations in many biological processes, e.g., auditory nerve firings, ion channel kinetics, foetal breathing, lung inflation, human cognition, blood pressure, heart rate, etc., even observed in human walking. Many other kinds of physical systems demonstrate the presence of fluctuations having a varied nature, as well as positive aspects.

In the investigation of the emergence and nature of noise, different mechanisms have been proposed, such as intermittency and self-organized criticality. But, the problem regarding the origin of noise still poses many puzzling questions needing to be answered in many branches of biological systems. Thus, taking into account the widespread nature of noise in complex and diverse biological systems, and, observing their fluctuating nature, it has been suggested that their noisy behavior can be explained by the fact that the final output could only be based on a system affected by the action of many processes on different time scales. Given this approach, it can be proposed that determining Fisher information content (Frieden 2004) could be relevant and hence applicable in neuronal communications. In mathematical statistics, **Fisher information** (sometimes, simply called **information**) can be defined as a method where the measurement of the amount of information that an unknown parameter carries can be obtained and what kind of information a random variable X carries about the observable upon which the probability of X depends. In explaining this problem formally, it can be stated as the variance of the score or it can be expressed as the expected value of the observed. In Bayesian statistics, the asymptotic distribution of the posterior mode depends on the Fisher information only, not on the prior distributions involved. It may be possible that the principle of least time and also that of the sum-over histories (Schuster 1904) can appear to be an important, viable principle in understanding the coherence dynamics, together with the degree of responsibility they possess for action and perception. In this way, the use of the principle of least-time, coherent noise cancellation combined with the intrinsic noise embedded in the signal, may be considered an essential one.

For more than 200 years, probability theory has been applied by mathematicians and philosophers in order to describe and analyze details about human cognition. At the beginning, this theory was developed basically for analyzing games of chance. Gradually, it has occupied a central role in formalizing various other kind of situations, i.e., how rational agents should reason in the case of uncertainty (Gigerenzer et al. 1991; Hacking 1975). The great potential of probabilistic models of cognition comes from the solutions they identify for inductive problems which play a central role in cognitive science. According to Griffiths et al. (2010), for most cognitive phenomena, including acquiring a language, a concept, or a causal model, some kind of uncertain conjecture from partial or noisy information is critically needed. This brings the requirement of employing a probabilistic framework to address a few key questions about these phenomena like:

- How much information is needed?
- Which kinds of representations preliminarily serve the inferences people make?
- What constraints are necessary on learning?

These are **computational-level** questions, and they are most naturally expected to be answered by **computational-level** theories.

In fact, the history regarding the probabilistic models of human cognition or thought is as old as probability theory itself (Chater et al. 2006; Oaksford and Chater 2007; Griffiths et al. 2010); Cosmides and Tooby 1996). Probability theory can be used both as a normative theory for proper reasoning about uncertain events as well as a descriptive theory about the reasoning of people regarding uncertainty. Bernoulli (1654–1705) in his famous book The Art of Conjecture, talked about how to embody this uncertainty, suggesting: "… both as how to guide for better reasoning and a survey how the art is actually practiced. That is, from its origin, probability theory was viewed as both mathematics and psychology" (Chatter et al. 2006).

The principle of Bayesian probabilistic rules of inference (Box and Tiao 1973) has been more and more applied to understand human learning and inference. This model has been increasingly studied over the broad spectrum of cognitive science, for example, in animal learning, human inductive learning, visual perception, sensory and motor control, language processing, symbolic reasoning, etc. There, Bayesian probability is used to make inferences about the world based on the observations as well as prior beliefs (Lasky and Cohen 1986). Feynman proposed (Feynman 1965) an alternative approach where inferences may be based on posterior probabilities. The Bayes factor can be considered as an example of this kind of approach. This approach, based on posterior model probabilities, may appear to be theoretically more appealing than that based on posterior parameter samples because the former approach appears much more difficult to implement than the latter. Fully general, and at the same time efficient, algorithms for computing posterior probabilities have yet to be developed. The above mentioned framework is different from other probabilistic approaches like the "frequentist" approach. Neyman stated that, in the case of frequentist probability (in a classical sense), it is necessary to consider random and repeated sample collections of trials. As is well known, the "frequentist" approach focuses mainly on summary statistics for the observed data (like the mean or the expected value of an observation), without considering the underlying causes that generated the data. According to this interpretation, probabilities are discussed only when dealing with well-defined experiments (or random samples). This approach to inference is based on the simple formula known as Bayes' rule (Bayes 1763/1958). Bayes inference and the details of probability theory will be discussed in next chapter.

Next the biggest challenge appears. Maybe we recall the famous quote of Terry Pratchett and Stephen Baxter in *The Long Earth*, before going into the details of the problem: "Maybe the only significant difference between a really smart simulation and a human being was the noise they made when you punched them". Let us put the questions as follows:

- How does the mind go beyond the data of experience?
- Realistically speaking, how does the mind build abstract, verifiable models of the external world based on the sparse and noisy data that we observe through our senses?

Before going into the details of modelling the noisy data or modelling uncertainty in cognition process, it is important to understand the meaning and role of noise, especially in brain function. As stated earlier, noise has generally been regarded as a **truly** fundamental engineering problem, particularly in electronics computation and communication sciences, where the aim or the requirement has been reliability and optimization. In non-linear dynamics, the discovery of stochastic resonance (SR) (Hanggi 2002) brought a transformation of that perception, i.e., noise, rather than representing a problem, became a central parameter in the system function, especially when considering cases in biology. Indeed, it can be stated that noise plays a basic role in the development and maintenance of the life, as a system, capable of evolution. But the question at this point is: *To what extent is biological function dependent on random noise?*

Thus, a crucial corollary to that question can be put regarding its viability and the degree of its applicability in biological phenomena. From various recently established findings, both from theoretical and experimental biological system research, it is now known that the detectability and transduction of signals are improved by the addition of the noise, especially in non-linear systems. This effect, popularly known as stochastic resonance (SR), has already been considered as an established phenomenon in sensory biology. But, presently, the question still remains about the degree of applicability, i.e., it has not yet been determined to what extent SR is embedded in such systems. As we are aware, SR, by nature, is a very conspicuous and ubiquitous phenomenon which suggests something about its careful application and corresponding interpretation in the neural models and noises of the brain function concerned. Thus, at the present stage, for instance, we are not yet sure what should be the proper amount and degree of the functionally significant SR, which can cause effects at the level of single-ion channels in cell membranes. Also, it is not yet decided whether it (SR) is mostly an ensemble property of channel aggregates or not. Currently, the presence of neuronal multi-modal aperiodic spike firing, in the absence of external input, has been found. From these examples, it appears that both forms of SR seem to be present, as it occurs based on external noise as well as in its absence.

It has been observed from both experimental findings and computer simulations of neuronal activity that SR is enhanced whenever both the noise and signals are simultaneously present in neurons. It is also known that spontaneous synchronization occurs if noise is added to the coupled arrays of neurons. Indeed, coherence resonance (CR) has been observed in the 'hippocampal' neuron array. It is to be noted that Derksen and Verveen (1966) changed the prevalent view of noise from considering it as a bothersome phenomenon and established it as an essential component of biological systems. From the recent developments in neuroscience, it has been pointed out further that the noise (carrying no signal information) can be utilized in the central nervous system (CNS) for enhancement of while signal detection. It should also be emphasized that, during information processing, a fundamental role of noise in the brain has been noticed. Beers et al. (2002), in their detailed review, discussed this problem with an example: whenever one tries to localize his hands, the CNS automatically integrates visual and proprioceptive information; not only that, these actions automatically possess different noise properties. It happens in such a way that, eventually, the ultimate result minimizes the uncertainty while estimating the overall. Secondly, they noticed that noise in motor commands always tends to 'leading to inaccurate movements'. But, when applied in the presence of signal-dependent noise to a framework of optimal-control, known as task optimization, it assumes the movements as automatically planned, and, by applying this technique, it can minimize the deleterious consequences of noise. As a final result, the inaccuracy developed due to noise can be minimized. The third point is that, during movement, sensory as well as motor signals need to be integrated in such a way, so that, finally, this will lead to the estimation of the body's state. According to Beers et al. (2002), "Neural signals are corrupted by noise and this places limits on information processing." They reviewed the complete processes related to goal-directed movements and observed what kind of determinative role is played by neural noise and uncertainty, which can influence many important aspects of our behavior. It is a well-known fact that noise limits the perception related to sensory signals. Different models are presented that show how these signals could be optimally combined. Finally, their work also explained how CNS deals with noise at the neuronal, including the network, level and carries out the tasks by minimizing the detrimental effects of noise. Moreover, the intrinsic noise, associated with neuronal assemblies, is known to produce synchronous oscillations which in turn, utilize the internal stochastic resonance (ISR) or coherence resonance (CR) mechanism.

The noise can be broadly classified as follows:

- **Basic physics noise**: Thermodynamics and quantum theory set physical limits on the efficiency of all information handling systems.
- **Stimulus noise**: Thermodynamics or quantum theory delineates the limits to the external stimuli and, thus, they are intrinsically noisy. During the process of perception, the stimulus energy is either converted directly to chemical energy (e.g., photoreception) or to mechanical energy, being amplified and finally transformed into electrical signals. The intrinsic noise present in the external stimuli will be amplified and further amplification of generated noise (transducer noise) will be produced.
- **Ionic channel noise**: voltage, ligand and metabolic activated channel noise
- **Cellular contractile and secretary noise**: muscles and glands
- **Macroscopic behavioral execution noise**.

There exist two distinct sources behind the execution of noise. They are:

1. **Non-linear dynamics**: where, in deterministic systems, the sensitivity to the initial conditions and to chaotic behavior engenders variability of initial conditions.
2. **Stochastic**: where irregular fluctuations or stochasticity may be present intrinsically or via the external world.

These two sources generate noise from chaotic time series or via stochastic processes. While they do share some indistinguishable properties, it is nevertheless possible to differentiate noise from chaos via stochastic processes. As previously mentioned, here also the important issues to be solved are: how and to what extent biological function is dependent on the presence of random noise, i.e., whether we can consider noise as a useful property in biological systems. The discovery of stochastic nature in non-linear dynamics addresses the above question directly.

Brain function deals with many vital and central issues in solving problems, for example, the internalization of the properties of the external world into an internal functional space. By the use of the term internalization, we mean the ability of the nervous system to fracture external reality, first into sets of sensory messages, and then to simulate such reality in the brain's reference frames. More than two decades ago, Pellionisz and Llinás (1982) proposed an integrated approach to address brain function. This was originally based on the assumption that the relationship between the brain and the external world is determined by the ability of the CNS to construct an internal model of the world implemented through the interactive relationship between sensory and motor expression. In this model, the evolutionary realm provided the backbone for the development of an internal functional geometry space. More recently, it was proposed that the CNS functional space is endowed with a stochastic metric tensor (Roy and Llinás 2008, 2012), such that a dynamic correspondence could exist between events in the external world with their corresponding specifications in the internal space. One can think of this type of correspondence between incoming (sensory) covariant vectors and outgoing (motor) action, i.e., contravariant vectors, as depending on internal states in a random fashion. We shall call this **dynamic geometry**. Here, dynamic geometry creates a state of readiness with respect to the external world, which is then used in a predictive anticipation of motor action. At the same time, this geometry also serves as an on-going contextual translator of covariant vectors into contravariant vectors. However, in the external world, the motion of an object does not itself engender simultaneity in space and time in its counterpart functional space in the brain, because conduction velocities are not constant among the various brain pathways, the delay due to deference in conduction velocities lying within 10–15 ms. The time resolution associated with gamma band (40 Hz) based cognitive resolution (10–15 ms) simultaneity may be considered, therefore, to have a granularity 'δt' (cognitive granularity) within that temporal framework (Kozima 2013). We will refer to this 'δt' as our operational definition of simultaneity. The variation of functional state generated here, by the summed activation in all pathways and corresponding neurons, in response to external or intrinsic activity, within this 'δt', will be referred to as a **quantum of cognition**. Instead of the usual optimization for the axonal pathway with maximum conduction velocity, a weighted summation over all pathways includes stationarity, which plays a preferential role in cognition and thus in our further understanding of the cognitive event.

This is comparable to the so-called sum over histories considered by Feynman et al. (1964, 1966) in connection with reflection and refraction of light. He emphasized these quantum principles and showed how, through the superposition of multiplications of unitary complex factors along all possible photons, they follow paths, resulting in light travelling between sources to detector in the least possible time. Thus for light, a sum of history results in a **least-time** displacement. By contrast, axonal paths with different conduction velocities can mimic light transmission in space only. When viewed from a morpho-functional perspective, informational conduction in the CNS is associated with distinct and particular conduction pathways with specific conduction properties. This way, in order to represent the external world in the brain, the set of different histories corresponding to different axonal paths sum up within individual neurons. We suggest that these conduction properties express a stochastic, a priori description of the world, consistent with the Bayesian perspective for brain function.

Anatomically, brain pathways connect functional spaces to one another. It is here then, that the sum of history principle and the resulting least-time principle, due to the oscillatory grains, can support quantum cognition by synchronization. We wish to emphasize that the principle of least time is a fundamental principle of nature that may help in defining the decision-making and behavioral control capacity of the brain. It is now clear from the above analysis that the ability to anticipate the outcome of a given action depends on the sensory stimuli from the outside world and the previously-learned experiencer's inherent instinct. But, the ability to predict the outcome of future events is, arguably, the most universal and significant of all global brain functions. Due to these reasons, it becomes essential to formulate a theory of inference using prior knowledge for decision making and judgment.

Typically, Bayesian models (Box and Tiao 1973) of inference are used to solve problems that involve probabilistic frameworks. Here, the term 'Bayesian inference' is used to denote all sorts of computations using Bayesian models, e.g., computing posterior probabilities and finding most probable explanations, as is common in the literature, and not only restricted in the manner of the formal *INFERENCE* problem in Bayesian networks. This is, of course, computationally interesting, as well as one of the most challenging aspects of cognition, and a variant of the classic problem of induction, as present in Western philosophy of knowledge including the philosophy of modern science. But it appears to be the core of the difficulties to be dealt with, when building machines with human-like intelligence. As expressed by Lisa Zyga (2014):

In quantum theory of cognition, memories are created by the act of remembering. The way that thoughts and memories arise from the physical material in our brains is one of the most complex questions in modern science. One important question in this area,–how individual thoughts and memories change over time? The classical intuitive view is that every thought or "cognitive variable" that we ever had can be assigned a specific, well-defined value at all times of our lives. But now psychologists are challenging that view by applying quantum probability theory to how memories change over time in our brains. Thus there are distinct two lines of thought when it comes to using quantum theory to describe cognitive processes.

Recently, through several experiments with human subjects (Aerts et al. 2013), it has been clearly revealed that traditional probability theory is violated in many cases and they suggested that, based on many literature surveys, classical probability theory, beyond a certain limit, is unable to model human cognition. The Bayesian approach may seem to be a promising candidate in dealing with this problem. However, the complete success story of Bayesian methodology is yet to be written. At any time, it appears to be a major problem to tackle the presence of epistemic uncertainty and its effect on cognition. Our established view is that the quantum view of uncertainty can basically be related to the uncertainty principle that is present in quantum mechanics. But a second school of thought proposes that the same can also be applied in the case of cognitive processes. According to their view, though basic physical processes in the brain are classical at the neuronal level, the apparent non-classical features of some decision-making arises because of the presence of the various complex ways by which thoughts and feelings can be related to those basic brain processes only. The cognitive version can be related to the entanglement of time. But again, to achieve this goal, it is necessary to have perfect knowledge of a cognitive variable at any point in time which, ultimately, will be able to bring the problem of some uncertainty into the process. It is worth mentioning that the cognitive domain stores thoughts over time and can be looked upon in the same way as we think of position and momentum in physics. In computational cognitive science, it appears that Bayesian computations can be modelled successfully for the applications to many cognitive processes. Yet, even when only an approximate solution is sought, for unconstrained input domains, many such Bayesian computations have been proved to be computationally intractable (NP-hard). This result of computational complexity appears to be in strong contrast to the ease and speed with which humans can typically make inferences, following Bayesian models. This contrast between theory and practice poses a great theoretical challenge for computational cognitive modellers: *How can intractable Bayesian computations be transformed into computationally plausible "approximate" models of human cognition?*

Three candidates for the notion of 'approximation' are discussed here. Each of them has been suggested in the cognitive science literature. John Kwisthout and Van Rooj (2013) have dealt in detail and discussed how (parameterized) computational complexity analyses can yield model variants that are tractable and which can serve as the basis of computationally plausible models of cognition. Moreover, the stochasticity in the model arises due to the unknown path or trajectory (definite state of mind at each time point) a person is following. Thoughts are linked in our cognitive domain over time. It is possible to consider these in the same way that position and momentum are linked in physics. To this end, a generalized version of probability theory, borrowing ideas from quantum mechanics may be a plausible approach. Quantum theory allows a person to be in an indefinite state (superposition state) at each moment of time. This characteristic allows all of these states to have a potential (probability amplitude) for being expressed at each moment (Heisenberg 1968). Thus a superposition state seems to provide a better representation of the conflict, ambiguity or uncertainty that a person experiences at each moment.

Conte et al. (2009: 1) demonstrated that mental states follow quantum mechanics during perception and cognition of ambiguous figures. They used ambiguous figures to analyze the existence of the violation of classical probability fields, if any, during perception-cognition of human subjects and to discover the existence of any possible role of quantum interference. The experiments were conducted on a group of 256 subjects. The results obtained brought them to the definite conclusion evidence that such quantum effects really play a role. Therefore, one of many propositions regarding the role of mental states, during perception-cognition of ambiguous figures, is considered to follow the principles of quantum mechanics. He also discussed in detail memory-related problems and the possibilities of using quantum theory in solving these problems of cognitive processes. The results can be well understood from his following statements of Conte et al. that summarize his findings:

> Another point of vital importance is the act of remembering which plays a vital role in quantum theory of cognition. Modern science faces one of the most grave and crucial question about tracing the ways, i.e., how the thoughts and memories arise from the physical material in our brains. These remain, still, one of the most complex questions in contemporary cognitive problems to be dealt with. The classical, institutive and conventional point of views proposes specific, well defined values, to every piece of thought or "cognitive variables" that we ever had, and also, can be assigned with a specific, well defined value of at all times of our lives. But now psychologists are challenging that view by applying quantum probability theory to how memories change over time in our brains. Thus there are distinct two lines of thought when it comes in using quantum theory to describe cognitive processes....

He further elaborated:

> Mental operations consist of content plus the awareness of such content. Consciousness is a system which observes itself. It evaluates itself being aware at the same time of doing so. We may indicate awareness statements by a, b, c, ... those are self-referential or auto referential and content statements of our experience by x, y, z, ... a = F(a, x) is the most simple definition of a single auto referential statement. For example, x = the snow is white; a = I am aware of this.

> Human experience unceasingly involves intrinsically mental and experiential functions such as "knowing" and "feeling", involving images, intentions, thoughts and beliefs. A continuous interface holds between mind/consciousness and brain. Neuroscience and

neuro-psychology have reached high levels of understanding and knowledge in this field by the extended utilization of electrophysiological and of functional brain imaging technology. Therefore, it becomes of fundamental and general interest for neuroscience and neuro-psychology to ascertain by experiments if quantum mechanics has a role in brain dynamics.

Another approach, known as Dempster-Shafer theory (Dempster 1967; Shafer 2008) that can handle epistemic uncertainty, still remains unexplored in the context of human cognition. This approach (as explained below) might explain the situations related to several uncertainties that traditional probability theory fails to explain. Here, one is allowed to make a decision at each point in time that may be distributed over an interval rather than conceived as a particular decision (confined at a point) leading to a possible avenue to tackle the so-called "interference effect" (Busemeyer et al. 2006; Spanos 1999) leading to a violation of traditional probability theory. Hence this theory should be investigated thoroughly in human cognitive domain. These studies clearly point to the two dominant approaches, i.e., the cognitive approach and psycho-biological approach to understand gambling behavior. The experiments based on gambling have been performed in the Western world, which differs culturally from the East. No similar experiments have yet been conducted in India or in any other eastern country. It is necessary to perform these kinds of gambling experiments in at least one Eastern country so that it could have been possible to compare the data with those were that collected in the West to identify any possible cultural dependency. These results would have more impact on resolving the issue of changing various paradigms than the classical probability theory (Einhorn and Hogarth 1981).

1.2 Emotion, Logic and Decision Making

A revolution in decision theory was launched by Herbert Simon (1967, 1983) who introduced the concept called 'bounded rationality' which requires refinement of existing models of rational choice to be considered in the case of cognitive and situational constraints. But, as he presaged regarding the necessity and validity of the role of emotions as connected to the incompleteness of decision theory, contemporary science, spanning from philosophy to neuroscience (Phelps et al., in press), points to the crucial role of emotions in judgement and decision-making (JDM) research. For many centuries and across various cultures, gambling has been treated as a form of entertainment. The studies on decision making during gambling (Clark 2010) raise lot of interesting questions regarding the role of the interaction between cognition and emotion. In fact, emotion has remained a black box in cognitive decision making for many years. The necessity of a theoretical model and its computational aspect is very much felt to answer the question "*how do emotions affect the cognitive function of a decision maker?*" One of the most challenging aspects of artificial intelligence (AI) is to model "human characteristics" like emotions, behavior, etc. in a comprehensive manner.

Over the last decades, there have been many theoretically revolutionary attempts that have yielded results with great potential to create a new paradigm for dealing with the decision-making problems closely associated with human characteristics, especially emotions. The factors associated with emotions assume the role of very powerful, pervasive and predictable drivers in nature, as during making decision. These phenomena can appear in diverse domains of decision making, influencing the regular, usual mechanism of decision processes, and thus influencing judgments and choices. As we are now aware, the decision mechanism is conduced through emotions and guides everyday attempts at avoiding negative feelings, e.g., guilt, fear, anger, etc., and, at the same time, maintaining positive ones (i.e., pride, happiness, love and compassion), even without being aware of those feelings. But, interestingly, this is not the end of story. Just after the materialization of these emotion-connected decisions, the feeling of new emotions, e.g., elation, surprise or regret etc. will arise in the picture.

Some attempts have been made during the last decade to develop a theoretical model regarding the role of emotion in decision making. Kim (2012) tried to develop a theory to consider "how the emotions affect the cognitive functioning of a strategic decision maker". Gershenson (1999) tried to model emotions with multidimensional logic, taking into consideration the role played by the degree of contradiction that the emotion might possess. The model is centered on simulating the role of emotions in artificial societies, generalizing the problem of contradictory goals and the conflicting goals of agents. A new surge of interest (Lerner et al. 2014) has arisen recently regarding the modelling of emotions in decision making and judgments. After 30 years of observations and data collections on emotion-based decisions and choices, Lerner and his group have developed an 'integrated model of decision making'. This incorporates a blend of both traditional (rational choice theory) inputs and emotional inputs, attempting a synthesis of both. In their pioneering paper, they surveyed the eight major thematic perspectives on the impact of emotions on judgment and decision making.

The present author emphasizes the use of quantum probability to model the decision-making process in the cognitive domain. However, quantum probability is an extension of quantum logic which is linked to density matrices in Hilbert space. Construction of Hilbert space in the functional states of neurons, or even in an abstract space of emotion, is a very difficult proposition from the biological point of view. On the other hand, the quantum analog of the total probability rule, within the framework of paraconsistent Bayesian probability theory (Salazar et al. 2014), is designed in such a way that it can handle the theory when inconsistencies or contradictions arise without leading to trivialization or logical explosion. To do so, it is required to assign probability to the occurrence of contradictions in paraconsistent Bayesian probability theory. These probabilities enter into the total probability rule depending on their values. Minsky (1985) discussed the possibility of machines having emotions and pointed out: "The question is not whether intelligent machines can have any emotions, but whether they can be intelligent without any emotions".

Here, a very vital point to be noted is to have a machine or computer capable of a simulation of emotions so that it will not only be intelligent, but also be 'believable". Here, the problem is that the emotions too often carry contradictions. Gershenson (1999) described the situation as:

> If your loved one cheats on you, we could say it makes you hate him or her. But that doesn't mean you stop loving him or her immediately. You feel love and hate for your loved one at the same time. Agreeing that love is opposite to hate, we find very hard to model this contradiction with traditional logic.

That is why we emphasize the use of paraconsistent logic. It is shown that quantum logic is a kind of paraconsistent logic. This will open up new vistas on the issue of affective and multimodal human–machine interaction.

We can visualize the organization of this book in the following manner: In the beginning, in Chap. 2, we discuss various approaches to decision making and judgment. Various types of models have been formulated for many decades in various contexts like social science, game theory, machine intelligence, computer science, statistics, etc. Since decision making in the real word is ultimately related to modelling in the cognitive domain, we will discuss all the approaches to decision making and judgment in the context of current understanding in neuroscience. However, the main goal of this book is not to discuss all the existing approaches within neuroscience and those in the context of modern neuroscience, but to discuss the various aspects of probability theory needed to understand the decision process from the point of cognitive neurodynamics. Among the various applications of probability theory, we discuss mainly the Bayesian approach and Dempstar-Shafer theory. These theories utilize traditional probability theory but deal with a more general framework. So, in Chap. 3, the necessity of Bayesian probability theory is discussed with the goal of understanding the functioning of the brain. The predictability of the brain is one of the major issues in modern neuroscience. This chapter deals with the recent findings of neuroscience, elaborated from the point of view of the neuronal circuitry responsible for predictability. This is closely related to the issue of decision making and judgment.

Then in Chap. 4, new empirical findings are discussed, classified into six categories. The standard probability theory is used to understand the data related to the evidence. It clearly shows the capabilities of the framework based on traditional probability (classical) theory to explain these data. Several authors have proposed that **"quantum probability"** is able to explain the results of the experiments performed for the six categories. The law of combining probabilities in classical probability theory fails to explain these results. On the contrary, interference effects, when applied in quantum probability, explain these results. Conte et al. (2009b) studied in detail the ambiguity of figures and claimed the inadequacy of classical probability theory. They performed an experiment searching for the presence of quantum mechanical effects. An important issue connected with that is to observe any possible manifestation, or the signature of the presence, of quantum interference effects. Here, during perception-cognition by human subjects, if any violation of the classical probability can be observed in the classical probability field,

a contribution from the role of quantum probabilities will automatically be established. Not only that, the observed results will be expected to differ from classical
ones, due to the presence of additional, quantum interference terms which will
appear as an obvious consequence. This situation could be expected, especially,
whenever ambiguous figures are to be analyzed. With this aim, as mentioned
earlier, when they conducted an experiment on a group of 256 subjects, they indeed
observed the presence of such quantum interference effects in their experiments.
Therefore, it should be emphasized that, during perception and cognition of
ambiguous figures, mental states appear to follow quantum mechanics.

However, it is necessary to study the role of non-commutability of the operators
so as to justify the quantum paradigm. It is worth mentioning that several authors
indicated the necessity of using non-commutability of the operators in the visual
domain so as to explain the ambiguity of figures. These results are addressed in a
separate section in this book so as to justify the use of the quantum paradigm, at
least in the visual domain (Gibson 1979). It is not yet clear whether the
Dempstar-Shafer theory can explain all of these results using the classical probability theory. The literature in this direction is scanty and more careful. More careful
and convincing investigations are necessary to justify the claim proposed by this
theory to serve as a means to incorporate the incompleteness of evidence into
simulation models. The key characteristics of this theory are that precision in inputs
is required only to the degree that can be justified by available evidence. Another
key feature of the Dempster-Shafer theory is the precision in input requirements
which, interestingly, is needed only to the degree justified by available evidence.
The output belief function consists of an explicit measure of the firmness of output
probabilities. This follows the belief function theory which presents its basic
methodology for application and corresponding simulation involving an example.
However, the theory of belief functions shows evidence that incorporates: *(i)* the
incompleteness of evidence in the simulation models; and *(ii)* the precision of
inputs, and the key feature of this theory is required only to a degree. This phenomenon can be justified by available evidence. The above mentioned belief
functions are called Bayesian functions, when defined probabilistically. A Bayesian
belief function, BEL (A) is defined as

$$BEL(A) = P(A|E)$$

where, E denotes the available evidence.

Pearl (1988) did makes these arguments with examples of how it is possible to
arrive at the Dempster-Shafer belief function, which ultimately leads to counterintuitive reasoning. He opined convincingly that, here, the probability can be
considered as a "faithful guardian of common sense". Again, Lindley (1987)
contends that the probability calculus is "the only satisfactory description of
uncertainty".

There have been two primary criticisms of probabilistic approaches as a representation of uncertainty. Firstly, there is no way to distinguish between complete
ignorance and complete uncertainty, which problem has largely been tackled with

the advent of Bayesian belief networks. These networks provide natural and concise graphical representations of probabilistic dependencies. Examples are incorporated how Dempster-Shafer belief functions can lead to counterintuitive reasoning. In short, there have been two primary criticisms of probabilistic approaches to representing uncertainty. The first one states that there is no way to distinguish between complete ignorance and complete uncertainty. The other criticism, which has historical precedence, points to the Bayesian formalism that the need to consider probability dependencies makes probability calculations unfeasible. But this problem has largely been tackled with the advent of Bayesian belief networks. The applications of this idea provide a natural and concise graphical representation of the dependencies. Thus, the incarnations of Bayesian belief networks seem to have been around for some time, introducing the concept with terms such as influence diagrams, knowledge maps, and causal diagrams. However, Pearl (1988) has been largely credited with standardizing and popularizing Bayesian belief network. With the pioneering work in his book, he opined convincingly that the probability can be considered the "faithful guardian of common sense". He introduced the idea of directed acyclic graph (DAG) and also the idea of minimal independency map (MAP). This requirement that a Bayesian belief network's underlying DAG should be a minimal independency map of the problem domain is what enables computations on belief networks to be greatly simplified.

Fundamental mathematics and idea of quantum mechanics are discussed in Chap. 5. In the beginning of the 20th century, physicists performed some experiments that clearly demonstrated the inadequacy of classical physics and the failure of law of combination of probabilities as proposed by Laplace (1774).

The crisis was resolved with the new paradigm called quantum mechanics. Quantum mechanics is considered as a non-classical probability calculus which is based on non-classical propositional logic. These non-classical propositions form a non-Boolean–non-distributive–ortho-complemented lattice. Intuitively, it could be stated that non-Boolean logic (Krifka 1990) is needed to understand the interference term in the double-slit experiment performed by physicists, which clearly demonstrate the dual aspects, i.e., the wave and particle nature of the entities at the microscopic scale. In their recent experiments, Arndt et al. (1999) observed and demonstrated the presence of de Broglie wave interference of C60 molecules by diffraction at a material-absorption grating. This C60 molecule is the most massive and complex object in which wave behavior has been observed. It is a particularly interesting fact that C60, by nature, is almost a classical body, because of its many excited internal degrees of freedom and their possible couplings with the environment. We emphasize that this type of non-Boolean logic, popularly known as quantum logic, is needed also to explain the results in the cognitive domain. The framework of quantum logic has a long history. The complementary principle, as proposed by Bohr, is discussed in Chap. 6, and the concepts of quantum probability, along with quantum logic are elaborated in Chap. 7.

At this stage of development, the present existing framework is an abstract framework and seems to be devoid of any material content, for example, the existence of elementary particles and their localization. It is an established fact that, in our physical world, the constants like the speed of light (c), the Planck constant (h) and gravitational constant (G) have definite numerical values which play important roles in dealing with modern-age problems in physics. The challenging issue in the context of quantum logic is how to make it context dependent. Though this framework of logic is abstract, it is applicable to any branch of knowledge, e.g., social science, biology, etc. But, yet, the question of context dependence in quantum logic is a pertinent question to be answered. In modern physics, several attempts have been made so as to embed the above mentioned fundamental constants in the framework of "quantum ontology".

Now the next important questions appear:

- What is the epistemological status of the laws of logic?
- What short of arguments are appropriate for criticizing purported principle of logic?

Recently, Mittelstaedt (2011) discussed the existence of a possible quantum ontology where no material content of the microscopic domain is used, rather, the language and logic is used for the description of microscopic entities. Then, based on this view, both classical ontology (i.e., the ontology of classical mechanics) and quantum ontology are investigated within this abstract framework. Since this framework is devoid of any material content, it can also be applied to any branch of knowledge. This approach will be discussed in detail in Chap. 8.

Haven and Khrennikov (2013) tried to formalize the term "contextuality" within the framework of a probabilistic formulation. It is known that a quantum—mechanical event description is context dependent. The role played by quantum logic (non-distributive) can be stated as a partial ordering of context, rather than being described as the ordering of quantum mechanical events. Moreover, the way this kind of quantum logic is displayed following quantum mechanics could also be explained as a general notion of contextuality, well explainable in ordinary language. Their basic idea is to consider a context C, which is a complex of conditions in physical or social science or of the biological domain and two dichotomous variables d and e. One may think of these two variables as random variables but in the conventional sense of Kolmogorov (1933/1950) probability, they are not measurable functions. Within context C, these variables have probabilities

$$P(d = +1|C), \ P(d = -1|C), \ P(e = +1|C), \ P(e = -1|C)$$

$P(a = +1|C)$ is the probability that $a = +1$ under context C.

They indicated that "[t]hey can be statistical probabilities obtained via frequencies of results of measurements as well as subjective probabilities i.e. a priori assignments for the values of d and e". In a series of recent papers, McCollum (1999) has studied quantum logic in the context of neuroscience, specifically for the sensory-motor system.

Theories of decision making and judgment from the neuroscience perspective are still in their infancy. A few attempts have been made in this direction.

McCollum made an attempt to understand decision making based on his framework of quantum logic in the context of nervous system, which is discussed in Chap. 9. Finally, the conclusive remarks and future directions of research are discussed in Chap. 10. Moreover, the epistemological issues related to decision making and judgment in the context of quantum probability are also discussed in this final chapter. Various schools of eastern philosophy, both orthodox and heterodox, have debated issues of decision making and judgment for many centuries. The mind and its various states are investigated within their frameworks in this pretext. The concept *"equanimity"* (*sthitaprajna*) has been extensively discussed in the Bhagavad Gita as well as in Buddhist philosophy. For example, the superposition of mental states like 'happiness' and 'unhappiness' can be conceptualized as the superposition of two state vectors in quantum mechanics. This concept is like a state of mind, called the neutral mind in Buddhist philosophy, which sheds new light on understanding decision making and judgment, independent of any bias.

References

Aerts, D., Broekaert, J., Gabora, L., & Sozzo, S. (2013). *Behavioral and Brain Sciences, 36*(03), 274-276.

Arndt, M., Nairz, O. & Vos-Andreae et al (1999). Nature, **401**,680-682.

Bayes, T. (1763/1958). Biometrica, **45**,296-315.

Beers R.J. van, Wolpart, D.M. & Haggart, P. (2002). Current Biology;12(100; 834-837.

Box, G.E.P. & Tiao, G.C. (1973). *Bayesian Inference in Statistical analysis.*Wiley; ISBN 0-471-57428-7.

Busemeyer, J. R., Wang, Z., & Townsend, J. T. (2006). *Journal of Mathematical Psychology, 50* (3), 220-241.

Chatter N., Tenenbaum J.B. et al. (2006). *Trends in Cognitive Science,* Elsevier.

Clark, L. (2010). Philos. Trans. R. Soc. London, B. Biol. Sci.; **365**;(1538):319-330.; ibid; (2013); Behavioral and Brain Sciences, p 181-204;
DOI 10.1017/S01405X12000477.

Conte, E., Khrennikov, A. Y., Todarello, O., Federici, A., Mendolicchio, L., & Zbilut, J. P. (2009). Mental states follow quantum mechanics during perception and cognition of ambiguous figures. *Open Systems & Information Dynamics, 16*(01), 85-100.

Cosmides, L. & Tooby, J. (1996). *Cognition,* Elsevier.; ibid; (2005); Evolutionary psychology, *"Conceptual foundations, D.M. Buss (ed.); Handbook of Evolutionary Psychology"*, New York, Wiley.

Dempster, P. (1967). Annals of mathematical Statistics, **38**, 325-339.

Derksen, H.E. & Verveen, A. (1966). Science,**151**;1388-1389.

Edwards Ward (1954), Psychological bulletin;**41**;380-417.

Einhorn, H.J. & Hogarth, M. (1981). Ann. Rev. Psychol.,**32**,53-88.

Feynman, R., Leighton, R.R. & Sands, M. (1964,1966). *The Feynman Lectures on Physics, 3 volumes*; ibid; Feynman, R. & Hibbs, A.R. (1965); *"Quantum Mechanics & Path Integrals";* McGraw-Hill, New York.

Frieden, B. Roy (1998, 2004). *Science from Fisher information; AUnification;* Cambridge University Press.

Gershenson Carlos (1999). Modelling Emotions with Multidimensional Logic; http://cogprints. org/1479/1/mdlemotions.html.

Gibson, J.J. (1979). *The ecological approach to visual perception;* (Houghton Mifflin).

Giang, P.H. & Shenoy, P.P. (2005). European Journal of Operational Research, **162**,(2),450-467.

Gigerenzer, G., Hoffrage, U., & Kleinbölting, H. (1991). *Psychological review*, 98(4), 506.

Griffiths,T.L., Kemp, C. & Tenenbaum,J.B. (2008). *Bayesian Models of Cognition* in R.Sun(ed.), Cambridge University;

Griffiths, T. L., Chater, N., Kemp, C., Perfors, A., & Tenenbaum, J. B. (2010). *Trends in cognitive sciences*, *14*(8), 357-364.

Hacking Ian, M. (1975). *The emergence of probability: A Philosophical Study of Early ideas about Probability,* Cambridge University Press;

Hanggi, P. (2002); A European Journal Chem Phys Chem of Chemical Physics and and Physical Chemistry *"Chem.phys.Chem" 3, pp. 285-290.*

Haven E. & Khrennikov A. (2013). Quantum Social Science. Cambridge University Press, p. 128.

Heisenberg, W. (1968). In *"Quantum Theoretical Reinterpretation of Kinetic and Mechanical Reactions"* Sources of quantum Mechanics(English translation); ed. B.L. Van der Waerden.

Horvitz, E. (1999). CHI'99. *Proceedings of the Sigchi conference on Human factors in Computing systems*, ACM, New York, NY, USA, 159-165

Kadane, J.R. & Larkey, P.D. (1982). Management Science, **28**(2) pp. 113-120.

Kahneman, D. (2003). American psychologist; **58**(9), 697-720

Kolmogorov, A.N (1933/1950). *Foundations of theory of probability,* N.Y., Chelsea Publ Co.

Kozima, H. (2013). *Japanese Psychol. Research, Special issue: Cognitive Sci. approach to Developmental disorders"*; H. Murohashi(ed.) **55**(2),168-174.

Krifka, M. (1990). *Papers from the 2nd Symposium on Logic & language,* Lászlo Kámán & Pólos (eds.); Budapest.

Kwisthout John & Van Rooij (2013). *Bridging the gap between theory and Practice of Approximate Baysian Inferences*: Cognitive System Research 09/2013.

Laplace: *"Memoire sur les Probabilities"* (1774). See discussion by E. C. Molina in *"Theory of Probability, Some Comments on Laplace's Theorie Analytique,"* Bulletin of the American Mathematical Society, ¥oi. 36; (June 1930); pp. 369-392.

Lasky, K.B. & Cohen, M.S. (1986). *"WSC'86: procd. of the 18th. Conference on Winter simulation"*; pp 440-444, ACM, Newyork, NY, USA.

Lerner S. Jennifer et al (2014). Emotion and Decision Making (16 June, 2014, Draft paper).

Lindley, D.V. (1987). The probability approach to the treatment of uncertainty; Statistical Science, **2**;(1), 3-44.

Lisa Zyga (2014) http://phys.org/news/2014-03-quantum-theory-cognition-memories.html

Manski, C. F. (1977). The structure of random utility models. *Theory and decision*, 8(3), 229-254.

Manski, C.F. & Lerman, S.R. (*1977-1988). Econometrica: Journal of the econometric society.*

McCollum, Gin (1999). Int.Journ.Phys., **38**(12),19991201,3253-3267.

Minsky, M.L. (1985). The society of Mind, Simon and Shuster, Old Tappan, NJ.

Mittelstaedt P. (2011). *Rational reconstruction of Modern Physics,* Springer.

Oaksford, M. & Chatter, N. (2007). *The probabilistic approach to Human reasoning"*; Oxford University press.

Pearl J. (1988). *Probabilistic reasoning in Intelligent Systems: Networks of Plausible Inference,* Morgan-Kaufmann, San Mateo.

Pellionisz A. & Llinás R. (1982). Neuroscience; **7**(12), 2949-7.

Roy, S. & Llinás, R. (2008). *Dynamic Geometry, Brain function modelling and Consciousness*; Eds. R. Banerjee & B.K. Chakravorty; Progress in Brain research; **168, 133**.

Roy, S. & Llinás, R. (2012). *Role of Noise in Brain function; Science; Image in action;* Eds. Z. Bertrand et al., World Scientific Publishers, Singapore; pp 34-44.

Roy,S. & Llinás, R. (2012). Metric *Tensor as degree of coherence in the dynamical organization of the central nervous system(CNS),* Procd. Mathematics of Distances & Applications; eds. by M.M. Deja, M. Petitzen, K. Markov) (ITHEA, Sophia), pp 169.

Salazar R., Jara-Figueroa C., and Delgado A. (2014). Analogue of the quantum total probability rule from Paraconsistent Bayesian probability theory; arXiv:1408.5308 v 1.

Shafer Glen A. (2008). A better interpretation of Probabilities and Dempster-Shafer degrees of Belief; www.glenshafer.com/p/assets/downloads/dsbetting.pdf

Schuster, A. (1904). *An Introduction to the Theory of Optics*, London: Edward Arnold.

Spanos, A. (1999). *Probability Theory and Statistical Inference,* Cambridge University press.

Wang, Y. (2003): Brain and Mind:A Transdisciplinary Journal of Neuroscience and Neurophilosophy, **4**(2), 151-167.

Wang Y. (2005a) Brain and Mind: A Transdisciplinary Journal of Neuroscience & Neurophilosophy;**4**(2), 199-213.

Wang Y. (2005b) In Proceedings of the 4th IEEE International Conference on Cognitive Informatics (ICCI'05) (294-300), IEEE CS Press, Irvin, California, USA, August; ibid; (2005b); A novel decision grid theory for dynamic decision-making. In Proceedings of the 4th IEEE International Conference on Cognitive Informatics (ICCI'05), (308-314), IEEE CS Press, Irvin, California,USA, August; ibid;

Wang, Y. (2007); The Int. 1 Journ. of Cognitive Informatics and Natural Intelligence (IJCINI), **1**(1), 1-271(1), Cognitive Informatics (ICCI'03) (pp. 93-97) London, UK.,73,33-36.

Wang, Y., & Wang, Y. (2004); IEEE Transactions on Systems, Man, and Cybernetics (C); **36**(2), 203-207.

Wang Y. and Ruhe G. (2007). Int. Journ. of Cognitive Informatics & Natural Intelligence, **1**(2), 73-85, 73.

Weisman K. & Moss F. (1995). Nature, 373 (6509): 33–6.

Chapter 2
Various Approaches to Decision Making

Life is the art of drawing sufficient conclusions from insufficient premises
—Samuel Butler (Telling it like it is, Paul Bowden, p. 88)

Abstract In general, two paradigms deal with the categorization of decision theories: the descriptive and normative theories. Descriptive theories are based on empirical observations and on experimental studies of choice behaviors. But the normative theories specifically assume a rational decision-maker who follows well-defined preferences of behaviors as well as obeys certain axioms of rational processes. The axiomatic approach plays an important role in formulating these theories of decision making. In this process, theories of decision making are often formulated in terms of deterministic axioms. But these axioms do not necessarily account for the stochastic variation that attends empirical data. Moreover, a rigorous description of the decision process is provided only through real/time perception. Then it is possible to avail the real-time decisions by repetitive application of the fundamental cognitive process. In such a situation, the Bayesian framework provides readily applicable statistical procedures where typical inference questions are addressed. This framework offers readily applicable statistical procedures, and it is possible to address many typical inference questions. But, in many cases, the applicability of algebraic axioms comes into question concerning viability, especially, when the application connected to empirical data arises. Again, the probabilistic approach to decision making needs to be investigated properly in order to study the empirical data. In such cases, where typical inference questions are addressed, the Bayesian framework provides readily applicable statistical procedures. Attempt has been made to study the various aspects of the Bayesian approach to analyze the observational data found in the new empirical findings on decision making.

Keywords Normative model · Canonical approach · Axiomatic approach · Bayesian approach · Bayes' rule · Dempster–Shafer theory

© Springer India 2016   23
S. Roy, *Decision Making and Modelling in Cognitive Science*,
DOI 10.1007/978-81-322-3622-1_2

In the traditional approach to decision making or judgment, a comparison is made between a decision and a judgment (that is, a decision or judgment about what to do), and a standard one or "benchmark". This leads to an evaluation of whether a particular decision or judgment is good or bad relative to the standard one. Decision-making behavior is present in almost every sphere of life and so is studied in many very different fields, from medicine and economics to psychology, with major contributions from mathematics, statistics, computer science, artificial intelligence and many other branches of the scientific–technical disciplines. However, conceptualization of decision making and the methods applicable for studying it with success vary greatly. This has resulted in fragmentation of this field. According to Wang, "Decision making is one of the 37 fundamental cognitive processes modelled in the Layered Reference Model of the Brain (LRMB)" , where each of the disciplines explores specifically the special aspect of decision making connected to a particular problem in that discipline. So, the basic problem for the conceptualization of decision making is to address what kind of methods are needed to be applied and followed successfully, to study it. These methods vary in substantial ways, from problem to problem, the results of which lead to the fragmentation of this field into many categories.

The criteria for the success of this theory cover the whole decision cycle. They consist of, e.g., the framing of a particular decision which has been reached based on beliefs, goals, etc. They should also include the background knowledge of the decision-maker regarding the formulation of the decision options. It establishes more preferences compared to others, together with making commitments which ultimately initiate the making of some new decisions. These decisions may lead, again, to some other newly developing indecision. The next step of the problem concerned necessitates in bringing steps where, in turn, it is needed to incorporate the reasons towards the consecutive cycle about previous decisions as well as the rationales for them. This process, ultimately may lead either to the revision or abandonment of commitments that already exist. In this way, the theory of decision making depends on the successful roles played by other high-level cognitive capabilities like problem solving, planning and collaborative decision making. Three domains are responsible for assessing the canonical approach: artificial intelligence, cognitive and neuropsychology and decision engineering. These standards can be achieved or provided in "normative models". 'Normative' means relating to a model or an ideal standard. In this approach, people are primarily concerned in making their decisions logically and rationally.

The normative process is typically contrasted with informative (referring to the standard descriptive, explanatory or positive content) data. This can be termed as supplemental information, for example, additional guidance, tutorials, history, supplemental recommendations and commentary, as well as the background development together with relationships with other elements. Informative data is not a basic requirement and doesn't compel compliance. They are valuable because of their sets of rules or axioms which are derived from utility theory in economics and probability theory. These can be used to test the predictions about human behaviors. Biases are studied if the behavior deviates from the predictions of normative models. There are other approaches to decision making than the normative approach, e.g., the rational theory of decision making.

Thus decision making is linked with a vast set of phenomena and processes (Rangel et al. 2008; Sanfey and Chang 2008). For example, all human beings when in normal, healthy, mental and physical states, try to make decisions resulting in their natural control of their actions: "I would rather prepare for the exam instead of joining that evening party", i.e., some kind of self-conscious control of decisions and actions comes into play. But voluntary control of actions can be realized only from the angle of pure reflexes. When hit by the doctor in the right spot, the patient's foot moves without having any intention. But regarding reflexes, it can happen due to earlier learned experiences which could be encountered in most common experiences. An example of such reflexive decision making, i.e., of taking action (controlling the speed or when to stop, etc.), is when you start to apply your car's break even before you become aware of your conscious decision, and it happens as fast as a reflex. The classic experiment in the research field of human volition was performed by Libet (1985a, b). However, after a modified and modern version of this experiment, the authors (Soon et al. 2008) sought to predict the final response outcome, based on the activity patterns observed via localized brain waves. They observed the activity patterns in the frontopolar cortex V, seconds ahead of the "urge to move", which enabled them to predict the final choice (Soon et al. 2008). While the latter result was above chance, however, one cannot arrive at a reliable prediction. Research advances of modern neuroscience lead us to classify the approaches to decision making as follow in the next section.

2.1 Decision Making (On the Basis of Human Decisions)

To start with, on the basis of human decisions, decision making can be categorized so:

I. Decision making, developed for money making, i.e.,

- Mathematical formalization, made in game theory.
- Phenomenological models of rationality.

II. Decision Making (with application from and of neuroscience):
These are models based on learning theories from artificial intelligence (AI) research, i.e.:

- Based on the findings of neural correlates of realistic models of decision making in the brain.
- Models arising from understanding the cellular basis of cognition.

2.1.1 Canonical Approach and Normative Models

In general, two paradigms can be stated as dealing with the categorization of decision theories into descriptive and normative theories. Descriptive theories are based on empirical observations and on experimental studies of choice behaviors. But the normative theories specifically assume a rational decision maker who follows well-defined preferences of behaviors, as well as obeys certain axioms of rational processes. Many researchers (e.g., Zachery et al. 1982) have proposed three components of decision making as the main contributors, identifying them as:

1. *decision situation*
2. *decision maker*
3. *decision process*

Even if different decision makers possess cognitive capacities varying greatly in degree, the human brain shares similar and recursive characteristics and mechanisms, with all the core cognitive processes being interconnected within the mechanism of the brain. The canonical approach to decision making can be assessed in three domains:

1. cognitive and neuropsychology
2. decision making in artificial intelligence
3. decision engineering

All these standards can be achieved or provided in "normative models". Normative, by definition, relates to an ideal standard of model, or is based on what is considered to be the normal or correct way of doing something. Typically, the informative data approach is in contrast to the normative approach. The informative data approach can be applicable and related to standardized positive contents. Here, the data could have the nature of a descriptive or explanatory nature, which can be, for example, supplemental information or recommendations, additional guidance, etc. This also includes data from the tutorials, or from background history, development, and relationships with other elements. It is interesting to note that informative data does not require and compel compliance. They are valuable because of their set of rules or axioms which are derived from utility theory in economics and probability theory. These can be used to test the predictions about human behaviors. Biases are studied if the behavior deviates from the predictions of normative models.

There are other approaches to decision making than normative models, which vary in the extent they correspond to observed choices. But the normative approach can be stated as a rational theory of decision making. However, there exist differences between these three functions—descriptive, normative and prescriptive when choosing models, where the evaluation criteria are interesting to study. For example, empirical validity, i.e., the extent to which they correspond with the observed data, determines the evaluation of descriptive models, whereas normative models are evaluated only when the theoretical adequacy, i.e., the degree to which they provide acceptable idealization or rational choices, is provided. Lastly, the ability in helping people to make better decisions, i.e., their pragmatic values, are deciding factors for the evaluation of the prescriptive models. However, difficulties are faced in defining as well as in evaluating all these three criteria, which could be encountered by the students of philosophy of science. It is also a fact, nevertheless, that the criteria are obviously different, thus indicating that the argument posed in the case of normative model might not be an argument against a descriptive model and vice versa.

Next, considering the property of stochastic dominance, this condition is considered as the cornerstone of rational choice, and any theory contrary to this can be regarded unsatisfactory from a normative standpoint. A descriptive theory, on the other hand, is expected to be responsible in accounting for the observed violations of stochastic dominance problems 2 and 8 as stated and explained in Tversky and Kahneman (1986). Only prescriptive analysis could be developed in order to eliminate and reduce such violations. The failure of dominance therefore could be the only answer to this kind of violation culminating in the proper counter-example to a normative model. It is due to this fact that this could be the observation to be explained by the descriptive model, and it appear to be a challenge for a prospective model.

2.1.2 The Axiomatic Approach

The axiomatic approach, belonging to the first category of approaches, plays an important role in formulating theories of decision making (Roberst 1979). In this process, theories of decision making are often formulated in terms of deterministic axioms that do not necessarily account for stochastic variations that attends empirical data. But, a rigorous description of the decision process is only possible when provided through real-time perception. Then it is possible to avail the real-time decisions by repetitive application of the fundamental cognitive process. In many cases, however, the applicability of algebraic axioms comes into question regarding viability, especially when applications connected to empirical data arise. The axiomatic approach is intended to characterize the fundamental principles of human decision making, which provides the necessary inputs to find out the essentially important and sufficient conditions needed for the existence of numerical representations. Usually, the deterministic axioms are the real base upon which the decision-making theories are formulated, and these axioms do not take into account the stochastic variation associated with empirical data. But the probabilistic approach to decision making, again, needs to be investigated properly to study the empirical data. In such cases, where typical inference questions are addressed, the Bayesian framework provides readily applicable statistical procedures. This situation arises when algebraic axioms are applied to empirical data, and the applicability of these axioms is to be tested. It is well known that employing a prior distribution is the key idea of the Bayesian framework where the parametric order constrains the implications of a given axiom. Here, the Bayesian framework is discussed, considering this as a suitable probabilistic approach to handle the empirical data employing a prior distribution that represents the parametric order constraints, implied by a given axiom. Usually, to estimate the posterior distribution, modern methods of Bayesian computation such as Markov chain Monte Carlo (McMC) is used. The advantage of following this method is that it provides the needed information enabling an axiom to be evaluated. Specifically, the descriptive adequacy of a given model is assessable only when we adopt the Bayesian p-value as the criterion. In turn, this can assess the descriptive adequacy of a given axiomatic model so that it becomes possible to select the deviance information criterion (DIC) among a set of candidate models.

Bayesian framework, thus, can be tested in this way and at the same time helps us to test already established, well-known axioms of decision making which also includes the axioms of monotonicity of joint receipt and stochastic transitivity. It is a well-known fact that monotonicity is a property of certain types of digital-to-analog-converter (DAC) circuits where the analog output always increases or remains constant whenever the digital input increases. This characteristic becomes an important factor in many communications applications in which DACs are used. It is interesting to note that such applications can function in the presence of nonlinearity, but not in the presence of non-monotonicity.

2.1.3 Bayesian Probabilistic Approach

Since this book studies modelling in the cognitive domain, specifically we will focus our attention on the Bayesian approach in the context of brain functions. Our thoughts, though abstract, are determined by the actions of specific neuronal circuits in our brains. The new field under consideration is known as "decision neuroscience" which is meant to discover these circuits, thereby mapping thinking on a cellular level. This way, cognitive neuroscience, a joint investigation with the aim of investigating the nature of human intelligence, is connected to decision making. Particular emphasis is given to improving these functions through cognitive neuroscience, based on the knowledge of the neural basis on which human beings make decisions. This includes two aspects, i.e., how we learn the value of good, as well as the actions to be taken.

Among all global brain functions, arguably, the ability to predict the outcome of future events is universally the most significant. The results of a given action depend not only on the ability to anticipate the amount of the outcome, but also on sensory stimuli. This information is fed by not only the outside world but also from previously learned experience or inherited instincts. As a next step, applying prior knowledge to decision making, as well as that to judgment, is then absolutely needed to develop a theory of inference. Typically, it has been found that the Bayesian models of inference are useful in solving such problems involve probabilistic frameworks. According to Atram (1985: 263–264):

By nature, human minds everywhere are endowed with common sense. They possess universal cognitive dispositions that determine a core of spontaneously formulated representations about the world. The world is basically represented in the same way in every culture. Core concepts and beliefs about the world are easily acquired, yet they are restricted to certain cognitive domains and are rather fixed. Not all advocates of domain specific biological constraints on cognition are so concerned to stress that only one outcome is permitted.

The basic question, addressed by the Bayesian framework is to find out the way how one updates beliefs and achieves the result, makes inferences, and applies the observed data. However, appropriate application of Bayesian analysis in the human cognitive domain has remained largely unexplored. Only a few attempts have been made (Griffith et al. 1998; Yu and Smith 2007; Berniker and Körding 2008; etc.) to model or explain human-brain dynamics rather than the cognition mechanism (although the two are definitely related). In this regard, Bayes paradigm may offer a possible solution by offering a tractable version of the so-called hierarchical modelling of the Bayes paradigm, which may be termed as the "Bayesian Brain Hypothesis" (Friston 2003, 2005; Knill and Pouget 2004; Doya et al. 2007). According to Griffith et al. (2000: 29):

Cognitive processes, including those which can be explained in evolutionary terms, are not "inherited" or produced in accordance with an inherited program. Instead, they are constructed in each generation through the interaction of a range of developmental resources. The attractors which emerge during development and explain robust and/or widespread outcomes are themselves constructed during the process. At no stage is there an explanatory stopping point where some resources control or program the rest of the developmental cascade. "Human nature" is a description of how things generally turn out, not an explanation of why they turn out that way. Finally, we suggest that what is distinctive about human development is its degree of reliance on external scaffolding.

Thus, this characteristic turns out to be a major virtue of the hierarchical, predictive coding account. Following Clark (2012), we can arrive at the conclusion that this method effectively implements a computationally tractable version of the so-called Bayesian Brain Hypothesis. *"But can Bayesian brains really be the same as a predictive brain? Or is the claim merely informal or imprecise, shorthand for something which is formally and factually false?"*

However, even besides being an intrinsically interesting and suggestive one, demonstration of behavior through this model needs to be established. Especially, the need resides in the observation of facts by which one could arrive at strong conclusions regarding the shape of the mechanisms that generates those behaviours. In the hierarchical predictive coding framework, the brain is assumed to represent the statistical structure of the world at different levels of abstraction by maintaining different causal models which are organized, accordingly, on different levels of a hierarchy. In such a case, each level of abstraction obtains input from its corresponding subordinate level. Now, predictions, for the level below, are made by employing a feed-backward chain. This means that the errors obtained between the models' predicted and the observed (for the lowest level) or inferred (for higher levels) input at that level, are used. These are carried out:

(a) In a feed-forward chain to estimate the causes at the level above and
(b) To reconfigure the causal models for future predictions.

Ultimately, this will produce the needed output for stabilizing the system when the overall prediction error is minimized. In his famous paper, Griffith (2000) beautifully expressed his views regarding the "developmental systems" perspective to replace the idea of a genetic program. This idea appears to be clearly and boldly convergent with the recent works in psychology involving situated/embodied cognition and the role played by, so to say, external 'scaffolding' in cognitive development. He stated as follows:

> Developmental Systems Theory (DST) is a general theoretical perspective on development, heredity and evolution. It is intended to facilitate the study of interactions between the many factors that influence development without reviving 'dichotomous' debates over nature or nurture, gene or environment, biology or culture. (Paul E. Griffith, Discussion: three ways to misunderstand developmental system theory, Biology and Philosophy (2005); 20; 417)

At this stage of development, we can now add, for example, the evolutionary accounts of art, religion and science (Mithen 1996), family dynamics, conscience (de Waal 1996), categorization (Atram 1990), cooperation (Sober and Wilson 1998) and cheating detection (Cosmides and Tooby 1992). On the other hand, many other researchers (Byrne and Whiten 1988; Whiten and Byrne 1997) proposed new theories of human cognitive evolution. Not only that, but a strident movement has been started, even in psychology and anthropology during the same period. The arguments made against earlier theorists in that discipline concerned neglecting the basic fact about the mind. Their proposal considered theories of human cognitive evolution as, basically, a product of evolution (Cosmides and Tooby 1992). This gives rise to the necessity of taking into consideration and paying attention to this context, and much research is yet to be accomplished. Thus, the very idea, which once we would have considered as a failure of rationality, now appears to be likely interpreted as good evolutionary design. Gigerenzer (2000) opined that, under the condition of our evolving statistical reasoning abilities, statistical information would accumulate in the form of natural frequencies. So, nothing appears wrong with our rationality if we like to use it in the domain for which it was designed, maybe, exclusively for this purpose. Currently, much attention has been focused on the suggestion that the evolved regularities in human cognition are very rich and domain specific. This fact has profound effects on the conceptualization of innate theories of the domain, such as grammar, physics, biology and psychology, to name a few of some of the most widely discussed spheres (Carey 1985a, b; Keil 1979; Welman 1990; Wellman and Gelman 1992; Gunner and Maratsos 1992; Pinker 1994). In fact, we do not construct these theories on the basis of the evidence available to us; instead, we inherit them from our ancestors. Such claims are commonly expressed in terms of biological "constraints" on the human mind and believed to be constructed in order to reason in particular ways.

Very recently, several theories of human memory have revolutionized our perspective on human cognition. They propose memory as a rational solution to computational problems posed, especially by the environment. The first rational model of memory offered the idea along with convincing demonstrations that human memory appears to be well adapted in a remarkable way to environmental statistics. But proper assumptions for these ideas are needed, at a bare minimum, as long as the form of environmental information is represented in memory. Several probabilistic methods have been developed recently for representing the latent semantic structure of language in order to delineate the connections to research in computer science, statistics and computational linguistics.

Now, before going into the details of the Bayesian model for cognitive science, let us elaborate the Bayesian framework itself. The Bayesian rule of decision theory is named after Thomas Bayes, who lived in the eighteenth century. After the death of Bayes, his friend Richard Price sent Bayes' papers to 'Philosophical Transactions of the Royal Society' for publication. In 1974, the paper *"Essay towards solving a problem in the doctrine of chances"* was published and became the mainstream of statistics. It is well known that among different branches of statistics, the two major paradigms in mathematical statistics are the frequentist and Bayesian statistics. Under the uncertainty of information contained in empirical data, methods based on the statistical inference and decision making are efficiently handled by Bayesian methods. Bayesian inference has tremendous impact on different kinds of popular components among formal models of human cognition (Chatter et al. 2006). This approach provides a general methodology which can be derived from an axiomatic system.

But Kwisthout (2011) raised some serious questions about a few difficulties while discussing the computational tractability of this method. They pointed out that it is necessary to focus on the cause estimation steps present in the feed-forward chain. They argued: instead one should question whether the predictive coding framework specifies satisfactorily the causes, given the necessary steps to be taken so that these steps can be both Bayesian and computationally tractable. In the Bayesian interpretation of predictive coding (Friston 2002), the estimation of the causes means finding the most probable causes ϑ_m, when the input μ are given for that level and the current model parameters θ, i.e.,

$$\Pr(\vartheta|u; \theta) = \arg \max \vartheta_m$$

Now, if ϑ_m input has a maximum a posteriori probability (MAP), the idea that Bayesian inference is implemented by the predictive coding appears to be primarily dependent on this step. Also, in addition to this, the idea that the predictive hierarchical coding, as a next step, makes Bayesian inference tractable is dependent on the presumed existence of a computational method that could be tractable. But, computing MAP—exactly or approximately—is already found to be computationally intractable because of the presence of causal structures (Shimony 1994; Abdelbar and Hedetniemi 1998). It is a well-known fact that the existence of tractable method is clearly dependent on the structural properties of the causal models. So, as opined by Blokpoel et al. (2012: 1):

At present, the hierarchical predictive coding framework does not make stringent commitments to the nature of the causal models, the brain can represent. Hence, contrary to
suggestions by Clark (2012), the framework does not have the virtue that it effectively
implements tractable Bayesian inference….. three mutually exclusive options remain open:
either predictive coding does not implement Bayesian inference, or it is not tractable, or the
theory of hierarchical predictive coding is enriched by specific assumptions about the
structure of the brain's causal models… assume that one is committed to the Bayesian
Brain Hypothesis, then the only third survives instead of first two options.

2.1.4   Bayesian Statistics

The basic idea of Bayesian statistics is that the probabilities in this framework are
interpreted as rational, conditional measures of uncertainty and loosely resemble the
word probability in ordinary language. Here, in various propositions, probability
represents the degree of belief whereas, the Bayes rule updates those beliefs. This is
based on some kind of new information and the strength of this belief which can be
represented by a real number lying between 0 and 1. Its other basis can be traced to
the idea of expressing uncertainty about the unknown state of nature in terms of
probability. Basic characteristics of this kind of statistics are that it is always
possible to update the obtained probability distribution in the light of the new data,
which solves many technical problems associated with standard statistics.

2.1.5   Bayes' Rule

Bayes' rule in probability theory and its various modes of applications can be stated
as follows: Let us consider two random variables A and B. Now, applying the
principle of probability theory, a joint probability of these two variables, $P(a, b)$ can be written by taking particular values of a and b for A and B respectively, as
the product of the conditional probability of $P(a)$ and $P(b)$, i.e., when A will take
on value a given B, and B takes on value b for given A, we have:

$$P(a, b) = P(a|b)P(b).$$

Using a symmetrical argument, i.e., without having preference of choosing
A over B in factorization of joint probability, $P(a, b)$ can be written as:

$$P(a, b) = P(b|a)P(a).$$

Rearranging the above two equations we have:

$$P(b|a) = P(a|b)P(b)/P(a).$$

This is known as Bayes' rule by which the conditional probability of b, i.e., $P(b|a)$ given a, can be obtained from the conditional probability of a, given b. It is important to note that in terms of random variables, this is nothing but the elementary result of probability theory. Usually, Bayes' rule can be thought of as a tool, updating our belief about a hypothesis A in the light of new evidence B. More technically expressed, our posterior belief $P(A|B)$ is calculated multiplying our prior belief $P(A)$ by the likelihood function $P(B|A)$ that B will occur if A is true.

For example, the observed data D can be analyzed within a statistical framework considering a formal probability model $\{p(D|\omega), \omega \in \Omega\}$ for some unknown value of ω over the parameter space Ω. According to the Bayesian approach, prior to the data being observed, it is necessary to assess a prior probability distribution $p(\omega|K)$ over the parameter space Ω, as it describes the available knowledge K about ω. Then the next step is to obtain posterior distribution from Bayes' theorem with probability density $p(\omega|D, A, K)$:

$$p(\omega|D,A,K) = p(D|\omega)p(\omega|K)/\int_{\Omega} p(D|\omega)p(\omega|K)d\omega$$

with A the assumptions made about the probability model. Here, all the available information about ω, after the data D is observed, is contained in the posterior distribution.

2.2 Decision Making and Statistical Inference

Statistical inference is a decision-making process under uncertainty that deduces the properties of an underlying distribution by analysis of data and draws conclusions, for example, about populations or scientific truths from the results. There are many modes of Bayesian and non-Bayesian approaches for performing statistical inferences which have been discussed by various authors in different contexts. This includes statistical modelling, data-oriented strategies and explicit use of designs and randomization in analyzes. Furthermore, there are broad theories (frequentists, Bayesian, likelihood, design-based, etc.) with numerous complexities (missing data, observed and unobserved confounding biases) for performing inference. At present, we focus our discussions on the Bayesian approach to decision making, especially in the cognitive domain.

Generally, when there are two or more possible courses of action, one faces a decision problem. Consider the class of possible actions be designated by A and for each, $a \in A$, let us denote the set of relevant events by Γ_a. By denoting the consequence of a chosen action as c $(a, \gamma) \in C$ where the event is denoted by $\gamma \in \Gamma_a$, then the decision problem can be described by the class of pairs $\{\Gamma_a, C_a\}$. Here, the possible actions are considered to be mutually exclusive. Then again, minimum collection of logical rules is required for "rational decision making". So as to avail a minimum collection of these types of logical rules, it is possible to propose different set of principles. For this reason, the rational decision making can be defined as a logical, multistep model for choice. It is expected to make a choice only between alternatives that will gain maximum benefits for themselves, but at the minimum cost. This follows an orderly path starting from problem identification through the expected solution.

In fact, it can be expressed in a straightforward way, i.e., this model not only assumes that the decision maker has full or perfect knowledge about the information regarding the alternatives supplied as a prior information, but also expects to have the time, cognitive ability, response and resources needed for the evaluation of each choice against the others. Basically, Bayesian decision theory is based on two basic tenets:

- Decision maker's choice or preferences are affected by new information through its effect on his rather beliefs than his taste.
- Decision maker's posterior beliefs represented by posterior probabilities are obtained by updating the prior probabilities (representing his prior beliefs) based on Bayes' theorem.

Probability theory has become the focus of attention only recently in cognitive science for the following reasons:

- In cognitive science, the focus was mainly on the computational aspect but not on the nature of inferences, probabilistic or not.
- For uncertain situations, such as in psychology and artificial intelligence, formal approaches are mainly based on non-probabilistic methods, for example, non-monotonic logic and heuristic techniques.
- The applications of probabilistic methods, in some cases, are considered to be too restrictive for the cognitive domain.

However, after the remarkable technical progress in mathematics and computer science using probabilistic models (Yuille and Kersten 2006), it has been possible to substantially reduce these restrictions. Tversky and Kahneman (1983) and their collaborators, in their classic works, suggested that human cognition is non-optimal, non-rational and non-probabilistic. But the answer depends on the definition of the word "rational". The cognitive scientists observe that we live in a world of uncertainty and that rational behavior depends on the ability to process information effectively despite various types of ambiguities. We investigate decision making in the domain of neuroscience in the next chapter. Presently, we focus our attention on understanding Bayesian probability theory in the domain of human cognition. As opined by Suedfeld (1992: 435), in his famous article "Cognitive managers and their critics":

The performance evaluation of decision-maker performance almost always results in the finding that leaders do not live up to the criteria of rationality and complex thinking espoused by the researcher. However, these criteria are not necessarily correct or even relevant. Decision-makers must cope with uncertain, ambiguous, changing, inadequate, and/or excessive information; high threat and reward; different time perspectives and pressures; and a multiplicity of values, goals, constraints and opportunities. As cognitive managers, they need to make good meta decisions (i.e., deciding what strategy to adopt and how much time and efforts to expend on particular decisions). No simple prescription, whether it advocates adherence to formal logic, understanding of the laws of probability, or maximal complexity of information search and processing, can adequately guide this effort.

Thus, in connection to the rationality of thinking and decision making, cognitive scientists observe that we live in a world of uncertainty. Rational behavior depends on the ability to process information effectively despite various types of ambiguity. But, then "are people rational?"

This is a complex question of "Action selection", quite a fundamental decision process for us on which the states, both of our body and the surrounding environment, depends. It is a well-known fact that the signals in our sensory and motor systems are corrupted by variability or noise. Our nervous system needs to estimate these states which depend on several factors including also the associated roles of these very factors. To select an optimal action, it is necessary to combine these state-estimates with knowledge of the potential costs or rewards of different action outcomes. Different mechanisms have been employed for studying the nervous system and the decision problems connected to this, specially, for the estimation followed by proper applications. It is interesting to note that the results obtained so far emphasize that human behavior is quite close to that predicted by Bayesian decision theory. This theory not only defines optimal behavior in a world characterized by uncertainty, but also describes coherently various aspects of sensorimotor processes.

Among many objectives meant for the nervous system, the central and primary one is to sense the state of the world around us and to affect this state in such a way so that it becomes more favorable and easy to us to deal with than the previous one. Sensory feedback is used for measuring this state. But, practically speaking, this information is available with questionable precision and subject to noise. Only limited levels of precision are available connected to our sensory modalities, which vary depending upon the situation. It is a quite well-known fact that vision becomes limited under dim light or extra-foveal conditions, hearing turns out to be unreliable for weak sounds, and proprioception drifts without calibration from vision (Hoyer et al. 2003; Ma 2008; Denève 2008; Beck et al. 2011; Zemel et al. 1998; Pouget et al. 2003).

Furthermore, our actions always cannot claim to have a deterministic outcome as these are subject to noise and uncertainty. Same negative comments could be made about our motor commands, which, in each stage, are associated with various kinds of noise, and even the properties of our muscles, themselves being subject to various kinds of fluctuations, vary day to day, not to mention the variations in strength, i.e., our bodily systems having varying degrees of fatigue linked with health, both physical and mental. Therefore, for successful achievements of its objectives, the nervous system must apply a method for the integration of sensory information into a cohesive whole and, at the same time, this must be done before choosing among actions leading to uncertain outcomes. Thus we arrive at the conclusive point when the nervous system faces a substantially crucial question, i.e., how the noisy information could be tackled so that only the positive contributions would be available and the negative avoided?

As we are already aware, a statistical framework is necessary for efficiently managing uncertain information. Bayesian integration is the mathematical framework by which uncertain information from multiple sources can be calculated and combined optimally. This framework can estimate the results in a coherent and maximally accurate way which it derives from a set of observations. Therefore, this method can be applied to integrate sensory information about the world and, then, to use this uncertain information to make choices about how to act accordingly in a fruitful manner. Above all, the Bayesian framework supplies us with a principled formalism with the help of which it is possible to track down the ways how an optimal nervous system senses its surrounding world and then acts upon it. To illustrate these ideas, let us consider the act of descending a staircase: as a first step to achieve this, the nervous system must sense the stairs, and then act accordingly for transporting us down them. Based on our familiarity with walking down stairs, we have preliminary but strong expectations for certain parameters like the distance between steps, their height and their general shape. Quite often these expectations are strong enough. As a result, when we descend stairs without looking at our feet or in the dark, we feel quite comfortable taking stairs without even observing them, though, normally, we first try to assess what is the proper distance we need to cover for a single step and other possible criteria to be taken care of, if necessary. In fact, vision does not provide perfect measurements, on the contrary, and provides us with an approximate estimate that might be very near to the actual measure of the step's height. Bayes' rule, on the other hand, defines how to combine our expectations of the step's height etc., without the visual sense which is expected to make an optimal estimate of the same. Related to another important point, an example could be put like this: the moment we start to take a step, we simultaneously receive sensory information about our on-going motion and also of height of the stairs. As a next step, we are able to combine this sensory information with the action we have just chosen that makes an optimal estimation of where we are and where are we headed. Finally, the Bayesian framework is utilized in choosing how we should step, given all our expectations and sensory estimates of the steps.

2.2.1 Bayesian Probability and Cognitive Domain

The Bayesian approach to inference has drawn much attention to the subject by the community working in the broad spectrum of cognitive science. During the last decade, many authors addressed issues like animal learning (Courville et al. 2006), motor control (Körding and Wolpert 2006a, b), visual perception (Yuille and

Kersten 2006), language processing (Xu and Tenenbaum 2008), semantic memory (Steyvers et al. 2006), language acquisition, etc., using Bayes' rule. Many such research programs have continued in order to study these aspects starting in the last decade or so. However, the great challenge is to learn about the mysteries of the human mind, which goes beyond the data experienced or, in other words, how the mind internalizes the external world out of the noisy data collected through the sensory organs. This is the most challenging aspect of cognition, not only from the point of view of computation, but also from the version of the age-old problem of induction in Western thought. The same problem also lies at the core of the debates that deal with building machines having human-like intelligence (robotics).

The various methods of probability theory not only characterize uncertainty in information but also manipulate the information itself. Not only that, this helps also for optimization. Bayesian rules, quite contrary to other variants of probability theory, help cognitive scientists to define the rules of rationality. This rule updates the belief of the agents from new data in the light of information as well as prior knowledge. As mentioned earlier, the uncertainties in the decision problem, considered to be unknown numerical quantities, can be represented, for example, by Γ (possibly a vector or matrix) possessed by the agents. Classical statistics are directed towards the use of sample information in making inferences about, say, Γ. These classical inferences are, for most of the cases, made without taking into consideration the use, i.e., for which purpose they are required. But in decision theory, on the other hand, an attempt is made to combine the sample information with other relevant aspects of the problem with the goal of arriving at the best decision. In addition to the sample information, two other types of information are typically needed, the first of which is the knowledge of possible consequences of the decision taken. Often this very prior knowledge can be quantified by determining the loss that could be incurred for each possible consequence of the decision. This loss, generally, can be quantified by determining the loss that would be incurred for each possible decision and for the various values of Γ. The Bayesian rule combines information in an optimal way, based on prior belief, with information from observational data. Within the framework of Bayesian decision theory, it helps during the choice of the action to maximize the performance related to particular tasks (Jacobs 1999, 2002). In Bayesian models, probability computations are applied to explain learning and reasoning, instead of hypothesis spaces of possible concepts, word meanings, or causal laws. The structure of the learners' hypothesis space reflects their domain-specific prior knowledge, while the nature of the probability computations depends on the domain-general statistical principles. Bayesian models of cognition thus combine both approaches which have historically been kept separate for their philosophical differences-providing a way to combine structured representations and domain-specific knowledge with domain-general statistical learning. Battaglia et al. (2003) made a comprehensive overview of Bayesian modelling and Bayesian networks. According to their findings, the use of sensory information is found to be satisfactorily efficient in making judgments and for the guidance of successful actions in the world. Added to this, it has been argued that the brain must represent and make use of the information gathered about uncertainty, both for perception and action in its computations. Applications of Bayesian methods, already a time-tested technique, have been successfully accomplished in building computational theories for perception, as well as for sensor motor control. Not only that, but sufficient evidence has been provided in the case of psychophysics which has been establishing examples and proof that 'Bayes' optimal' is the only answer to human perceptual computations.

With these specific characteristics, the 'Bayesian coding hypothesis' states that the brain represents sensory information probabilistically, i.e., in the form of probability distributions. Many of the proposals deal with several kinds of computational schemes. This type of approach, of course, puts special emphasis on the viability and degree of success connected to these schemes, with the aim of making their model successful in dealing with populations of neurons. Neurophysiological data on this hypothesis, however, is almost non-existent. Due to this insufficiency of data available to neuroscientists, this situation poses a major challenge to test these ideas experimentally, i.e., how to determine through which possible process neurons code information about sensory uncertainty will be successfully applied. They mainly focused on three types of information processing. Specifically, these are:

- Inference
- Parameter learning
- Structure learning.

These three types of operations are discussed in the context of Bayesian networks and human cognition. These types of Bayesian networks become more and more popular in the field of artificial intelligence and human cognition since the factorization of a joint distribution is expressed by graph theory where a network contains nodes, edges and probability distributions. The models, developed following a Bayesian framework, do not follow the usual algorithm or process level. On the other hand, this characterizes more the usual, traditional cognitive modelling, but in the spirit of *"Marr's computational theory"*.

Marr is best known for his pioneering work on vision. But, before starting that work, in his three seminal publications, he proposed computational theories of the cerebellum (1969), neocortex (1970), and hippocampus (1971). In each of these mind-boggling papers, crucial: he presented new ideas, which continue to influence modern theoretical thinking to the present day. His 'cerebellum theory' was motivated by two unique features of cerebellar anatomy:

1. The cerebellum contains vast numbers of tiny granule cells, each receiving only a few inputs from "mossy fibers";
2. Among Purkinje cells, present in the cerebellar cortex, each receives tens of thousands of inputs from "parallel fibers", but, surprisingly, only one input from a single "climbing fiber" is extremely strong.

According to Marr's proposal, the granule cells encode combinations of mossy-fiber inputs and the climbing fibers carry "teaching" signals. This signal instructs their corresponding Purkinje cell targets to modify the strength of their synaptic connections with parallel fibers. Though neither of these ideas is universally accepted, yet both of them form the essential elements of viable modern theories.

To be very precise, Hubel and Wiesel (1974) found several types of "feature detectors" in the primary visual area of the cortex, and these primarily motivated Marr to introduce the theory of the neocortex. Generalizing those observations, he proposed that cells in the neocortex are a kind of flexible categorizer, i.e., they learn, primarily, the statistical structure of their input patterns and become sensitive to frequently repeated combinations. Again, the theory of the hippocampus (named "archicortex" by Marr) was developed from their discovery by Scoville and Milner (1957). The latter demonstrated that destruction of the hippocampus produce amnesia but only for the new memories or recent events, while this damage leaves the memories of earlier events that occurred years ago intact. It is interesting to note that Marr described his theory as a phenomena of "simple memory" only: The basic idea behind it was that the hippocampus, by strengthening connections between neurons, could rapidly form a simple type of memory traces. Marr (1982) emphasized that, for carrying out the information processing tasks, vision could very well be considered as responsible. Not only that, but any such task, he argued, could be described in three levels:

(i)   computational theory
(ii) specific algorithms, and
(iii) physical implementation.

These three levels correspond roughly to:

(i)   not only defining the problem but also setting out the way for solving the problem, at least minimally, so that, in principle, it can be solved
(ii) designing a detailed simulation of the process and be available at hand and
(iii) building a working system for carrying out the proposed problem.

The important point to be noted here is that, in this problem, the levels can be considered independently. As a result, it should be possible to mimic the algorithms underlying biological visions in robots: the only difficulty to be overcome is to find out the possible and viable ways for the physical implementation of these criteria. This concept of independent levels of explanation remains as a "mantra" in vision research even today. Marr made a serious attempt to set out a computational theory for vision in its totality, emphasizing that the visual process passes through a series of processes, where each corresponds to a different representation, say, from retinal image to '3D model' representation of objects.

Remarkably, Marr's paper preceded only by two years a paper by Bliss and Lømo (1973) that provided 'the first clear report' about the long-term potentiation in the hippocampus, a type of synaptic plasticity, very similar to what Marr hypothesized. This vital observation was reported in Marr's paper also as a footnote that mentioned a preliminary report of that discovery. Though Marr's theory regarding his understanding about basic concept of hippocampal anatomy has turned out to contain errors and hence has become less important, nobody can deny his work regarding the basic concept of the hippocampus as a temporary memory system, which remains a part of a number of modern theories (Willshaw and Buckingham 1990). At the end of his paper, Marr promised a follow-up paper on the relations between the hippocampus and neocortex, but no such paper ever appeared. However, some phenomena are described in a more satisfactory way at the algorithmic level or at the neuro-computational level. Here, it is important to mention that all the models of human cognition based on different levels of computation are not applicable to Bayesian analysis. In fact, problems related to inductive inference help fruitfully to solve the problems related to Bayesian rules, and in a more natural manner.

A great deal of theoretical and experimental work has been done in computer science, on inductive inference systems, i.e., the systems that try to infer general rules from examples. But, it still remains a far-reaching goal to suggest a successful and applicable theory for such a system. To start with, it is necessary to survey highlights and explain the main ideas that have already been developed in the study of inductive inference. At the same time, special emphasis should be given on the relationships between the general theory and the specific algorithms and implementations. This is needed for surveying the essential characteristics, both the positive and difficulties related to the techniques which have already been developed. However, the Bayesian approach is becoming more and more relevant and successful in many areas of cognitive science, whereas, previously these areas were conceived in terms of traditional statistical inference (Anderson 1990). Many perceptual phenomena can be explained parsimoniously using a Bayesian approach (Knill and Richards 1996). Bayesian inference fits well with all of Marr's levels of description. It is a useful tool in describing a problem at the level of computational theory, especially as adopted in current biology. A Bayesian model of a motion illusion occurs, i.e., when a narrow, low-contrast rhombus, is moved to the right—it appears to move down as well. This can be understood by: (i) considering the set of stimuli that could have produced the edge-motion signals the observer receives; and (ii) including a 'prior' assumption that objects tend to move slowly (adapted from Weiss et al. 2002). Weiss et al. (2002: 598) formulated a model of visual perception, applying standard estimation theory in which they stated:

> The pattern of local image velocities on the retina encodes important environmental information. Although humans are generally able to extract this information, they can easily be deceived into seeing incorrect velocities.... (i) There is noise in the initial measurements and (ii) slower motions are more likely to occur than faster ones. We found that specific instantiation of such a velocity estimator can account for a wide variety of psychophysical phenomena.

According to this approach, these velocities would, eventually, all fall along a line for a high-contrast edge. The 'aperture problem' is considered to be behind this fact which makes it impossible to measure exactly the velocity of an edge. The cause behind this is that no local motion of the signal movement in the direction of the edge is produced. The line appears blurred for the low-contrast stimulus.

Because, moving at other velocities, edges produce the same motion signals, having some noise in the very same system. Barlow (2001) reviewed this problem in detail where he discussed Marr's notion of independence between levels, together with the theories of neural architecture in the brain that might carry out this kind of inference. According to his views, this technique can be developed to deal with the generic quantities without reference to specific stimuli (reviewed by Barlow 2001). Gibson (1979) advocated that Bayesian approaches also demonstrated their effectiveness in the results when applied not only to the evolutionary but also to the ecological perspective. He advanced the theory of direct perception and proposed that perception and action are one and the same thing. Not only that, but an observer and the environment can be considered as an inseparable pair, which, as per his views, in principle, uses real-life stimuli under natural environmental conditions. The core concepts to be taken into consideration in dealing with such problems are, among others: invariance, affordances, and pi-numbers. According to Gibson (1979: 235), a few preliminary observations in ecological psychology can be stated as follows:

> Prolonged distortion leads to recovery. When one puts on distorting spectacles, e.g., so that straight lines get curved, the visual system adapts to this. After some time, a few hours or days, the lines are seen as straight again. Perception is not the sum of simple sensations. How an observer recognizes visual objects such as human faces or animals cannot be derived from research on the perception of simple points and lines. In laboratory conditions, stimuli are impoverished and poor in information and the percept represent marginal phenomena. Perception must be studied in real-life conditions, not in laboratory conditions. In real-life conditions perceptual information is rich in structure and, hence, in information.

As per his views, the basic need of a simple organism with a simple behavioral repertoire is only to divide information about the organism's state into a small number of categories, basically with respect to the world. Following this technique, the simple organism concerned can use its motor system while moving between these categories (this, however, remains the sole way by which it can get the knowledge about the success of its motor movement). But a greater number of states are required for reliable discrimination whenever dealing with the case of a more complex behavioral repertoire. In generating different motor outputs, this requirement becomes essential for the sensory systems so that it can evolve as per the requirement which helps an organism in discriminating between the contexts. At various stages of a task, the sensory parameters become most helpful in discriminating, as well as controlling movements, quite differently. Again, from Gibson and his collaborators (Clutton-Brock et al. 1979):

> This leads to a view which states light enters the visual system as an optic array with highly complex, but structured, and rich in information. Moreover, by moving around in the environment, the flow of information over the senses is considered to be the essential source of information for the organism. The organism scans the environment in the course of perceiving. Hence, the observer and the environment are an inseparable pair; and perception and action cannot be separated from each other (Clutton-Brock et al. 1979).

This leads to a view which states that the cortex is a pool from which evidence can be drawn. Thus, according to the demands of the task, in each consecutive moment, the neurons loaded with the most relevant information may be located in quite different parts of the cortex. Rizzolatti and Craighero (2004) proposed a theory based on their experimental observations in relation to the role of a category of stimuli. This appears of great importance for primates. They opined after their observations:

> Humans in particular, formed by actions done by other individuals ... to survive, must understand the actions of others...without action understanding, social organization is impossible. In the case of humans, there is another faculty that depends on the observation of others' actions: imitation learning. Unlike most species, we are able to learn by imitation, and this faculty is at the basis of human culture ... on a neurophysiological mechanism–the mirror-neuron mechanism ... play a fundamental role in both action understanding and imitation... we describe first the functional properties of mirror neurons in monkeys... characteristics of the mirror-neuron system in humans... those properties specific to the human mirror-neuron system that might explain the human capacity to learn by imitation.

2.3 Dempster–Shafer Theory

The Dempster–Shafer theory (DST) is basically a theory of belief functions, also referred to as evidence theory. It is a general framework which deals with uncertainty in reasoning and has understandable connections to many other frameworks, e.g., possibility and probability, including imprecise probability theories. This framework can be generalized as a mathematical theory of evidence (Dempster 1967; Shafer 1976) for tackling problems connected to uncertain information. Dempster–Shafer (DS) structure (Shafer 1976; Klir and Yuan 1995) on the real line, though quite similar to a discrete distribution, is different for the characteristic which requires the locations at which the probability mass is assigned, rather than precise points. These are sets of real numbers, termed as 'focal elements'. Each focal element has a non-negative mass assigned to it. The basic probability assignment is zero but corresponds to the probability masses with these focal elements. The basic probability assignment can be related to the correspondence of probability masses associated with focal elements. However, in DS structure, the focal elements may overlap one another rather than be concentrated at distinct mass points, as found in a conventional discrete probability distribution. The idea here is to develop a belief function and a plausibility function based on either the sample observations and/or other evidence from prior knowledge. These beliefs and plausibility are expected to serve as lower and upper bounds respectively, with respect to the actual probability of occurrence meant for a particular event. Here, although DS structure exploits traditional probability theory, it gives more general structure to the uncertainty underlying the phenomenon. It is well known that a decision is affected by many such factors, i.e., information of both objective as well as subjective nature. Various models have been developed to handle objective and subjective information, such as interval numbers (Deng et al. 2011; Kang et al. 2011; Xu et al. 2006); fuzzy set theory (Deng et al. 2010; Ngai and Wat 2005) Dempster–Shafer theory of evidence, and so on.

In the decision-making process, at first, the objective information is collected and then the decision makers join their subjective preferences with the objective information to reach a decision. It is not always easy to combine the objective information with the subjective, e.g., in political decision making to select the optimal economic policy in a country. In decision making based on Dempster–Shafer theory (DST), the information is usually assumed to be exact numbers. In real situations, the information is usually imprecise or vague. DST is used based on the interval of numbers. Recently, a new approach to DST has been developed by Merigo, Ramon and many others Merigo and Casanova 2008; Merigo et al. 2009; Merigo and Engmenn 2010) to handle uncertain information using the method of uncertain-induced-aggregation operators. Both DST of belief functions and Bayesian probability theory (BPT) are two distinct frameworks dealing with uncertain domains. Although they have important differences due to their underlying structures in semantics, representations and the rules for combining and marginalizing representations, they have significant similarities, too. Cobb et al. (2006) discussed in detail about the differences and similarities between these two frameworks and finally came to a conclusion, stating that the two frameworks have "*Roughly the same expressive power*". The extra advantage of DST is that it needs weaker conditions than BPT. The belief function model or DST is shown to be a generalization of the Bayesian model (Shafer and Srivastava 1990). In addition, under an uncertain environment, DS theory has the distinct advantage of being a theory of reasoning. It is able to express the "uncertainty" by assigning the probability to the subsets of the set composed of **N** objects, instead of assigning it to each of the individual objects. Furthermore, this theory has the ability of combining pairs of bodies of evidence or belief functions to derive a new evidence or belief function.

Recently, Fox et al. (1993, 2003, 2005, 2010) proposed a unified approach to decision making called "a canonical theory of decision making" where they address the following questions:

- How can we understand the dynamic lifecycle of decision making from the situations and events that make a decision necessary, to influence their prior knowledge, beliefs, and goals which determine how a decision will be framed, preferences arrived at, and commitments to actions made (Fox and Das 2000)?
- What are the general functions that underpin and constrain the processes that implement such a lifecycle for any kind of cognitive agent, whether the agent is natural or artificial?
- How does decision making, conceived in this very general way, fit within cognitive science's strategic objective of having a unified theory of cognition (UTC) that can cut across psychology, computer science, artificial intelligence (AI) and neuroscience (e.g., Newell 1990; Anderson 2007; Shallice and Cooper 2011)?

- How can we apply this understanding to decision engineering, drawing on insights into how decisions are and/or ought to be made to inform the design of autonomous cognitive agents and decision support systems (e.g., Fox et al. 2003, 2010; Lepora et al. 2010)?

So, this appears to be a very ambitious theory. According to Fox et al. (2010), it is necessary to establish a framework where discussions between decision researchers in various communities are possible. This theory, however, advocates an interdisciplinary approach and needs more careful and analytic work to make it comprehensive.

2.3.1   Cognition and Emotion in Human Decision Making

Since in this book, among other things, we wish to draw the attention of scholars to this very intriguing but interesting line of research, i.e., to the new empirical evidence related to gambling and decision making, it is worthwhile to consider the *interaction between cognition and emotion in human decision making*. Much research of varied natures, i.e., investigations underlying biological, psychological or social factors, have been ongoing that are hypothesized to contribute to gambling behavior.

In recent decades, the gradual expansion of the availability of gambling facilities, specifically, in the most of affluent Western countries, are much more common than those available in most parts of the East. Due to this, there is a necessity for, and considerable interest in, conducting research, especially, in a field involving those people who develop problematic levels of gambling. In order to address this complex social and psychological problem, a large body of research has been conducted in order to understand the determinants of gambling behavior. Evidence now exists that biological, psychological and social factors are all interlinked for the development of problematic levels of gambling. However, the theoretical explanation for gambling has lagged behind the advances in empirical works until now. Clark (2010, 2012), discussed the two dominant approaches to gambling behavior: **cognitive** and **psychological**. A number of erroneous beliefs has been identified in cognitive approaches whereas case-control differences between groups of 'pathological gamblers' and 'healthy control' have been identified with the psychological approach. In short, impaired decision making is responsible and a key feature among many with neuropsychiatric disorders.

Franken et al. (2007), in their book "*Impulsivity is Associated with Behavioural Decision-making Deficits*" concluded, as a result of his experimental observations, that impulsivity and behavioral decision-making deficits are always associated with each other. Not only that, but impaired decision making is the origin and is responsible as key features of many neuropsychiatric disorders. In their experiment, they noted also the task performances in a healthy population whose scores indicated high and low impulsivity depending on different kinds of behavioral decision-making tasks. This reflected orbitofrontal functioning measures "included tasks" that assess decision making with and without a learning component and choice flexibility. The results in their experiments on decision-making performance pointed to the fact that subjects, specifically those exhibiting high impulsivity, display an overall deficit, showing weaknesses in learning reward and punishment associations. Not only that, but they are prone to be wrong in making appropriate decisions (e.g., in the reversal-learning and Iowa gambling tasks where it is, absolutely needed). Together with these observations, it has also been pointed out that the impaired adaptation of choice behavior is present, according to changes in stimulus-reward contingencies (reversal-learning task). On the other hand, the simple, non-learning components of decision making based on 'reward and punishment' (Rogers decision-making task) appears to be relatively unaffected. Their results indicate that, in response to fluctuations in reward contingency, the impulsivity observed is associated with a decreased ability in altering the choice of behavior. Furthermore, these findings also establish evidence that supports the notion that trait 'impulsivity' is associated with decision making, a function of the 'orbitofrontal cortex' only. The impulsivity is known to be associated with behavioral decision-making deficits followed by impaired decision making. This is considered to be a key feature of many neuropsychiatric disorders. So, the cultural, as well as biological dependence are to be taken into account, especially, in the case of gambling data where human decision making and judgment are critical.

Another typical gambling disorder (GD) has been noted in the case of those gamblers who face problems in stopping their gambling behavior once the process gets started. The hypothesis behind this behavioral patterns states that, on a neuropsychological level, it is the result of the cognitive inflexibility of pathological gamblers, which reflects in their impaired performance on neuro-cognitive inflexibility, and is task measured and reward-based. Their suggestion is that the cognitive inflexibility in GD is the result of an aberrant reward-based learning, and, 'not based' on a more general problem with cognitive flexibility. Boog et al. (2014), based on their detailed experiments on this problem, observed flexibility and the pattern of problems and suggested that GD is the result of an aberrant reward-based learning, not based on a more general problem connected to cognitive flexibility. These results indicate the cause behind GD to be due to the dysfunction of the orbitofrontal cortex, the ventrolateral prefrontal cortex, and also the ventral regions of the striatum in gamblers. It is quite well known how people make decisions, keeping in their mind varying levels of probability and risks associated with the steps of actions taken. But, as stated earlier, even in the present state of research, we are not that much aware of the missing information, not yet available. For this reason, little is known of the neural basis of decision making. Due to this prime ambiguity, as a consequence, the probabilities are also uncertain. In decision theory, ambiguity about probabilities is not at all welcome as it affects substantially the choices. Using functional brain imaging, Hsu et al. (2005) pointed out the role of different parts of the brain in decision making and have shown that the level of ambiguity in choice correlates positively with activation in the amygdala and orbitofrontal cortex but negatively with the striate system. They found positive correlation of striate activities with expected rewards. Not only that, but it has been observed that neurological subjects with orbitofrontal lesions were insensitive to the level of ambiguity and risks in behavioral choices. Finally, they suggested based on their observations that the response of a general neural circuit is dependent on the different degrees of uncertainty, contrary to decision theory.

Next, it is important to mention the vital role of noise, its imminent effects on nervous system and consequently the outcome regarding decision making (Körding and Wolpert 2006a, b). Let us explain the scenario in the following way: if a steady rate is considered an important criterion for determining the intensity of a constant sensory or motor signal, then any variation in the inter-spike intervals will cause fluctuations in the rate, which will appear as unwanted noise. But, if the timing of each spike carries extra information, then this very variability turns out to be a crucial part of the signal. We argue here that both temporal and rate coding are used, to varying degrees, in different parts of the nervous system, and this enables the nervous system to discriminate between complex objects. As a result, graceful movements are generated.

Nearly 60 years ago, it was Claude Shannon (see Gold and Shaden 2001), who developed a theory regarding the information that can be transmitted in a noisy communication channel. This theory has been found to be very effective for applications in many branches of science, especially computer science. In computers, this theory has become popularized in terms of the 'bits and bytes' of information and has even been applied to the information content of the universe (Noel-Smith 1958). Application of this theory to neural systems appears to be quite logical. In a noisy or uncertain environment, the survival of an organism is dependent on how rapidly it is able to collect the necessary crucial information. However, it is not at all clear whether information theory proposed by Shannon is applicable to biological system (Gatlin 1972). Gatlin addressed this problem beautifully in her book where she treated this problem from the angle of classical information theory but with different interpretations. She applied the concepts of information theory developed by Shannon and others to the living system in the process. But the information theory and the second law of thermodynamics are redefined and ingeniously extended in the context of the concept of entropy. Finally, let us conclude with Körding and Wolpert's remarks in this regard:

> Action selection is a fundamental decision process for us, and depends on the state of both our body and the environment. Because signals in our sensory and motor systems are corrupted by variability or noise, the nervous system needs to estimate these states. To select an optimal action these state estimates need to be combined with knowledge of the potential costs or rewards of different action outcomes. We review recent studies that have investigated the mechanisms used by the nervous system to solve such estimation and decision problems, which show that human behavior is close to that predicted by Bayesian Decision Theory. This theory defines optimal behavior in a world characterized by uncertainty, and provides a coherent way of describing sensory- motor processes (Körding and Wolpert 2006a, b: 319).

References

Abdelbar, A.M. & Hedetniemi S. (1998); Artificial Intelligence; 102(1), 21-38.

Anderson, John R. (1990); *"Cognitive Psychology and its implications (3rd edition)"*: W.H. Freeman; NewYork (series of books on Psychology).

Anderson, John (2007); *"How Can the Human Mind Occur in the Physical Universe"*? Oxford University Press.

Atram, Scott (1985); American Anthropologist, New Series; **87**,298-310; ibid; Atram, Scott (1990); *"Cognitive Foundations of Natural History: Towards an Antropology of Science"*: Cambridge University Press.

Barlow, Jessica A. (2001); Journ. Child. Lang.; **28**, 201-324.

Barry R Cobb, Prakash P Shenoy & Rafael Rumi (2006); Statistics and Computing; **16**; 293-308.

Battaglia P.W, Jacobs R.A, et al (2003); J. Opt. Soc. Am. A., **20**, 1391–1397.

Berniker, M. & Körding, K. (2008); Nature Neuroscience, **11**(12): 1454-61.

Beck, J.M., Latham, M., Latham, P.E., & Pouget, A. (2011): *J. Neurosci.* **31**:15310-15319 and references therein.

Bliss T., & Lømo T. (1973) *Journ. Physiol* **232**(2):331–56.—PMC 1350458. PMID4727084.

Blokpoel Mark, Johan Kwisthout and Iris van Rooij (2012) Frontiers in Psychology **3**, 1

Boog, M, Höppner, P. et al (2014) Front. In Human. Neuroscience; **8**, 1.

Byrne, R. & Whiten, A. (1988) *Machiavellian intelligence: Social expertise and the evolution of intellect in monkeys, apes and humans*: Oxford University Press.

Carey, Susan (1985a, b);(a) *Are children different kind of thinkers than adults"*; In S. Chapman, J. Segal & R. Glaser (Eds); *Thinking and Learning skills*; **2**, 485-518; Hilside, NJ, Erlbaum. (b)*"Conceptual change in childhood"*; Cambridge; MA, MIT press

Chatter, N, Tenenbaum, J.B. & Yuille, A. (2006); Trends in Cognitive Sciences; **10**(7), 287-291.

Clark, L. (2010); Philos. Trans. R. Soc. London, B. Biol. Sci.; **365**; (1538):319-330.

Clark, L., Studer, B. et al (2012); Frontier in Neuroscience (decision Neuroscience);**6**,46.

Clutton-Brock, S.D. Albon, R.M. Gibson. (1979). Animal behavior, **27**(1), 211–225.

Courville, A.C., Daw, N.D. & Touretzky, D.S. (2006) Trends in Cognitive Sciences; **10**, 294-300.

Cosmides, L & Tooby, J. (1992); *Cognitive adaptations for social exchange;* In J. Barkow, L. Cosmides, & J. Tooby (Eds.); *"The adapted mind"*, New York: Oxford University Press.

Dempster, A. (1967) *Annals of Mathematics and Statistics*; **38**, 325-33.

Deng, X, Zhang, Y. et al. (2010) *Journ. Inform. & Comput. Science*; **7**, 2207-2215.

Denève, S., *Bayesian Spiking Neurons I: Inference*, Neural Computation, 20, 91-117 (2008);ibid; *Bayesian Spiking Neurons II: Learning*, Neural Computation, 20, 118-145 (2008).

Deng, Y., J., Wu, J.X., et al (2011) ICIC Express Letters, **5**, 1057-1062.

Doya, K. et al (2007) *Bayesian Brain: Probabilistic approach to Neuoral Coding*; The MIT Press; edition (Jan 1).

F.B.M. de Waal (1996); Good Natured; The Origins of Right and wrong in Humans and other Animals: Harvard University press.

Friston, K.J., Glaser, D.E. et al (2002); NeuroImage, **16**:484-512.

Fox E., Hix, D., Nowell, L.T. et al; (1993); Journal of the American Society for Information Science, **44**(8): 480–491.

Fox J. & Das S.K. (2000); Safe and Sound; Artificial intelligence in hazardous applications: American association for Artificial intelligence, The MIT press, Cambridge MA, USA.

Fox, J., Beveridge,M., et al (2003) AI Commun. **16**, 139–152.

Fox, E., Karana, M., et al (2005); Acm *Transactions On Information Systems (*Tois*)*, 23(2): 147–168.

Fox, C.W., Billington, P., Paulo, et al (2010): *"Origin-Destination Analysis on the London Orbital Motorway Automated Number Plate Recognition Network"*; Euro. Transport Conference.

Franken, I H A. et al. et al (2007); Alcoholism-Clinical & Experimental Research, **31**, 919-927. doi: http://dx.doi.org/10.1111/j.1530-0277.2007.00424.x

Friston, K.J, Holmes, A.P. et al. (2003); Dynamic causal modeling; Neuroscience, **19** (04,1273-302.

Friston, K.J., (2005); Philos. Trans. R. Soc., London, B. Biol. Sci.; **360**(1456),815-36.

Gatlin Lila L. (1972); *Information Theory and the Living System*; Columbia University Press, NY, & London.

Gibson, J.J. (1979); The ecological approach to visual perception (Houghton Mifflin).

Gigerenzer, G. (2000); *Adaptive thinking, Rationality in the real world*: New York; Oxford University Press.

Gold, Joshua I and Shaden N. Michael (2001) Trends in Cognitive science; 5(1); **10**-16.

Griffith, T.D., Rees, G et al (1998); Nature.com; in Percept Psychology, **45**, 371-327.

Griffith Paul & Stotz K. (2000); *Synthese*; **(122)**, 29-51.

Gunner, M.R. & Maratsos, M. (1992); in *Language and Modularity and constraints cognition*: The Minnesota symposium on Child Psychology; 25, Hillside, NJ, Lawrence, Erlbaum associates.

Hoyer, P.O. & Hyvarinen, A. et al. (2003); Neuroimage; **19** (2), 253-260.

Hubel, D. H. & Wiesel, T. N. (1974); *Brain Research*; Elsevier.

Jacob, R.A. (1999); Vision. Res., **39**, 3621-3629;

Jacobs, R.A. (2002). Trends. In Cogn. Sci.; **6**: 345-50.

Kang, B., Zhang, Y.X. et al (2011) Journ. Inform. & Comp. Sci.; **8**, 842-849.

Keil, F (1979); *Semantics and conceptual development*; Cambridge, MA, Harvard University Press.

Klir, G.J. & Yuan, B. (1995); Proc. Ninth ACM-SIAM Symp. on Discrete Algorithm, 668-677.

Knill, D.C & Pouget, A. (2004) Trends *in Neurosciences;* **27** pp. 712-719.

Knill, D. & Richards, W. (1996) *Perception as Bayesian Inference;* Cambridge University Press.

Körding, K.P. & Wolpert, DM (2006); *Probabilistic mechanisms in sensor motor control*; *Novartis Found Symp.,* **270**:191-8; discussion: 198-202, 232-7.

Körding KP, Wolpert DM (2006) Trends Cogn. Sci, **10**(7):319-26.

Kwisthout, J. (2011) Int. Journ. Approx. Reasoning, **52**, 1452-1469.

Lepora, N., Evans, M., Fox, C.W., et al (2010); *Naive Bayes texture classification applied to whisker data from a moving robot:* Proc. Int. Joint/Conf. Neural Networks (IJCNN).

Libet, B. (1985a) Behavioral and Brain Sciences **8**; pp. 529-66.

Libet, B. (1985b) Journal of Theoretical Biology **114**; pp. 563-70.

Marr David, (1969) *J. Physiol.,* **202**:437–470;

Marr David (1970 *Proceedings of the Royal Society of London B",* **176**:161–234;

Marr David 1971) *Phil. Trans. Royal Soc. London",* **262**:23–81;

Marr David (1982) *Proceedings of the Royal Society of London B",* **214**:501–524. Markov, N.T, Ercsey - Ravasz, M.M. et al (2014); Cerebral Cortex, **24**:17 -36.

Merigo, J.M. & Casanova, M. (2008); *"Procd.IMPU'08";* (eds.; L. Magdalena, M. Ojedo-Aciego, J.L. Verdegay), 425-432.

Merigo, J.M. & Gil-Lafuente, et al (2009*); "Intelligence System Design & Applications"*; ISDA'09.

Merigo, J.M. & Engmann, K.J. (2010); *Procd. World Congress on Engineering*, **1**, WCE2010, June 30-July 2, London, UK.

Mithen Steven (1996) *The Prehistory of Mind: Cognitive Origins of Arts, religion and Science*; NewYork: Thomas and Hudson.

Ming, Hsu, Bhatt, Meghana, etal (2005); Science; **310** (5754),1680-1683.

Newell, Allan (1990) *United theory of cognition;* Cambridge, MA, Harvard University press.

Ngai, E.W.T. & Wat, F.K.T. (2005); Fuzzy decision support system for risk analysis in e-commerce development, Decision Support Systems; **40**; 235-255.

Noel-Smith, P.H. (1958); Choosing, deciding and doing; Analysis; **18,** 63–69.

Pinker, S. 1994. *The language instinct;* NY: Morrow.

Pouget A., Dayan, P. & Zemel, R.S. (2003); Annual Review of Neuroscience, **26**; 381-410.

Rangel, A., Camerer, C. & Montague, R. (2008); Nature Reviews Neuroscience; **9**; 545-556.

Rizzolatti G. & Laila, Craighero (2004); Ann. Rev. Neuroscience; **27**; 169-172.

Roberts, F. S. (1979); *Measurement theory with applications to decision making, utility and the social sciences*; London, UK: Addison-Wesley.

Sanfey A.G. & Chang L.J. (2008) *Annals of the New York Academy of Science*; **1128**(1)**;** 53-62.

Scoville, William Beecher & Milner, Brenda (1957) Journal of Neurology and Psychiatry; **20** (1):11–21.

Shallice, Tim & Cooper, Rick (2011) *The Organization of Mind*: OUP, Oxford (ISBN(s); 9780199579242).

Shafer, G. (1976) *A Mathematical Theory of Evidence;* Princeton University Press, Princeton.

Shafer, G. & Srivastava, R.P. (1990) A Journal of Practice and Theory, **9** (Suppl.), 110-48.

Shimony, S.E. (1994) Artificial Intelligence; **68**; 399-410.

Sober Elliot & Wilson David Sloan (1998); *Unto Others: The Evolution and Psychology of Unselfish Behaviour*; Cambridge, MA; Harvard University Press.

Soon, C.S., Brass, M. Heinze, H.J. et al (2008); Nature Neurosccience; **11**; 543-545.

Steyvers, M., Griffiths, T.L. & Dennis, S (2006): TRENDS in Cognitive Sciences, **10**(7), 327-334.

Tversky A. & Kahneman, D. (1983) Psychological Review, **90**(4); 293-315.

Tversky A. & Kahneman, D. (1986); The journal of Business; **59**(4); S251-S278

Weiss, Y., Simoncelli, E.P. & Adelson, E.H. (2002); Motion illusions as optimal percepts; Nat. Neurosci; **5;** 598–604.

Welman, H.M. (1990); *The Child's theory of Mind*; Cambridge; M A; MIT Press; A Bradford book.

Wellman Henry, M & Gelman, Susan (1992) Annual review of Psychology, 43, 337-375.

Willshaw DJ, Buckingham JT (August 1990); *Philos. Trans. R. Soc. Lond., B, Biol. Sci.* 329 (1253): 205–15.

Zachery, W., Wherry, R. et al (1982); Decision situation, decision process and functions: Towards a theory –based framework for decision-aid design; Procd. 1882 Conf. on Human factors in Comp. System.

Whiten, A. & Byrne, R.W. (1997); *Machiavellian Intelligence II : Extensions and Evaluations (Eds.)*; Cambridge University Press.

Xu, D. L. et al (2006); European Journal of Operational Research; **174;** 1914- 194.

Xu, F., & Tenenbaum, J. B. (2008); *Word learning as Bayesian inference; Psychological Review;* (The Cambridge Handbook of Computational Psychology, Ron Sun (ed.); Cambridge University Press.

Yu Chen & Lind Smith (2007); Cognition; **106**(3); 1558-1568.

Yuille, A.L. & Kersten, D. (2006); Trends in Cognitive Neuroscience; **10**(7); 301-308

Zemel, R.S., Dayan, P. & Pouget, A. (1998); Neural Computation; **10**; 403-430.

Chapter 3
Predictability of Brain and Decision Making

The question that immediately arises is whether the biological Phenomena themselves dictate or justify the theory's mathematical structures. The alternative is that the beauty, versatility, and power of mathematical approach may have led its aficionado to find areas of application in the spirit of the proverbial small boy with a hammer, who discovers an entire world in need of pounding.

—Perkel (1990)

Abstract The central issue in brain function is to understand how the internalization of the properties of the external world is realized within an internal functional space. By the term "internalization" of the properties, we mean the ability of the nervous system to fracture external reality into sets of sensory messages; the next task of this process is to simulate such reality in brain reference frames. As stated earlier, the concept of functional geometry has been proposed by Pellionisz and Llinas, and then developed further by Roy and Llinas as probabilistic dynamic geometry to understand the internal functional space and its correspondence with the external world. A central challenge concerning present-day neuroscience is that of understanding the rules for embedding of 'universals' into intrinsic functional space. The arguments go as follows: "if functional space is endowed with stochastic metric tensor properties, then there will be a dynamic correspondence between the events in the external world and their specification in the internal space." The predictability of the brain is closely related to dynamic geometry where Bayesian probability has been shown to be necessary for understanding this predictability. In Bayesian statistics, prior knowledge about the system is assumed. The fundamental question regarding the CNS function is "where is the origin of such prior knowledge", and the answer is basically simple: the morpho-functional brain network is initially inherited and then honed by use. In fact, the origin of a Bayesian framework can be traced to Helmholtz's (1867) idea of perception. It was Helmholtz who first realized that retinal images are ambiguous and prior knowledge is necessary to account for visual perception.

Keywords Dynamic geometry · Predictability · Functional space · Internalisation · Bayesian probability

© Springer India 2016
S. Roy, *Decision Making and Modelling in Cognitive Science*,
DOI 10.1007/978-81-322-3622-1_3

Let us start with three apparently distinct queries about brain functions (Roy and Llinas 2008) that can be stated as:

- The unknown related to how the brain represents the world and how it performs transformations on this representation
- The unknown related to sensorimotor coordination
- The unknown related to the physical basis of perceptual organization.

The internalization of the properties of the external world is the central issue in brain research. According to the beautiful Benjamin Jowett translation, we find that Plato (427–347 B.C.), in his well-known allegory of the prisoners in the cave (Plato, *The Allegory of the Cave*), discussed these kind of issues long before the development of modern brain research. As per Wittgenstein (Klagge 1999), a distinction can be made in between the world as the domain of our experience and the world as the domain of things present within it. So, the central issue in brain function turns into the topic where we need to understand how the internalization of the properties of the external world is realized within an internal functional space. Here, by the term "internalization of the properties", we mean the ability of the nervous system to fracture external reality into sets of sensory messages, the next task of which process is to simulate such reality in brain reference frames.

At this juncture, automatically a few of the so-called trivial questions come to mind:

> How is 'real' represented in our heads? Does cognition work with information from the 'outside' world? Is our thinking exclusively determined by the world? Is it exposed to perturbations from an apparently endless environment, or does cognition actively generate and construct this 'world'? (Markus et al. 1999).

In 1987, Rodolfo Llinás proposed that the brain (CNS) should be considered as a fundamentally closed system that is perforated and modulated by the senses. This is a basic issue that needs to be investigated with great rigor. The concept of open and closed systems has been discussed for many decades in the physical sciences. Developments in the study of thermodynamics have helped substantially in formalizing conceptual structures for open and closed systems. The work of Prigogine and Kondepudi (1998) took a further step in this direction. In physics, systems may be generally classified into the following three categories:

- Isolated system: There is no transfer of matter or energy between the system and the environment.
- Closed system: There is no transfer of matter but an exchange of energy is allowed and the system is open with respect to energy.
- Open system: This allows the transfer of matter as well as energy.

The issue of an open or closed system for biological systems becomes one of the central issues in twentieth century biophysics. Broadly speaking, an open system in physics is defined as a system that exchanges both energy and mass with its outside surroundings, its environment. In a more general context, an open system accepts external interactions, i.e., accepts input from external sources and produces output, defined as the case in which matter and/or energy may enter and exit, such as a human body. So, following the general approach, this can be considered as valid in any case of interactions, for example, any kind of information, energy or material transfers going into or out of the system boundary. But then, in the case of a biological system, it accepts inputs from the environment, processes them and returns them to the external world as a reflex, and consequently it is considered an open system. This, of course, depends on the discipline defining the associated concept. But an open system, also known as a constant volume system, or a flow system, is contrasted with another way of conceptualization, i.e., the concept of an isolated system.

A closed system, on the contrary, allows nothing to enter or escape its boundaries. This system, by definition, is not supposed to exchange any information with its environment, whether it is in the form of energy, matter or information. With such more extendable applications possibilities, the concept of an open system is applied extensively with the advent of information theory and, subsequently, system theory. Due to the broader aspects of approach in the system, thus, this idea has been introduced with success to various applications, e.g., natural sciences dealing specifically with energy and mass, as well as the social sciences. In the natural sciences, an open system is one whose border of application is permeable to both energy and mass. But, in physics, a closed system, by contrast, is permeable to energy but not to matter. Due to these above-mentioned characteristics, open systems have supplies of energy that cannot be depleted; are considered to be supplied from some surrounding environment; and have the possibility to be treated as a source with infinite capacity for the purpose of the study. Following this definition, an open system is supposed to interact with what is around it, and can be understood only by including an understanding of their relationship to everything else. In another way, we can say that it means—in a closed-system sense—that they can never be fully understood. That is why the biological system, which accepts inputs from the environment, processes them and returns them to the external world as a reflex, is considered to be an open system.

We consider the central nervous system (CNS) as a closed system in the sense that its basic organization generates intrinsic images such as thoughts or predictions, whereas inputs specify the internal states rather than "homuncular vernacles" (*homunculi* are "little men" in the brain). So, the CNS possesses autonomy, in the sense quite similar to that proposed by Maturana (2000). Thus, in order to realize what kind of autonomy exists in the crucial organizational properties, we need to understand that our main objective is to understand how a system can be related with the cognitive system of the world. Now, autonomous processes can also be called composed systems, which are capable of both the generation and sustenance of that very system as a unity and define also an environment related to the system. We can formally or concretely define autonomy, with respect to some abstract characteristics or in terms of its energetic and thermodynamic requirements

(Thompson 2007). Considering abstractly, for a system to be autonomous, its constituent processes must meet the following conditions:

- recursively depend on each other for their generation and realization as a network;
- constitute the system as a unity, in whatever domain they exist; and
- determine a domain of possible interactions with the associated world.

This definition basically follows Varela's (1979, 1997) propositions about the crucial property of an autonomous system regarding its *operational closure*. According to these propositions, every constituent process is conditioned by some other processes in an autonomous system. So, when analyzed for the enabling conditions, i.e., for any constituent process, the system leads to other possible kind of processes in the system, too. It is interesting to note that operational closure does not imply that the conditions "not belonging to the system" need to be declared as unnecessary. In contrast, to physically realize any autonomous organization, the operationally closed network of processes must be thermodynamically open. This leads us to the second way of characterizing autonomy. Our basic aim is to specify the energetic and thermodynamic requirements for the instantiation of "basic autonomy" in the physical world (Ruiz-Mirazo and Moreno 2004). From this perspective, basic autonomy is "the capacity of a system to manage the flow of matter and energy through itself so that it can, at the same time, regulate, modify and control:

- internal self-constructive processes and
- processes of exchange with the environment" (Ruiz-Mirazo and Moreno (2004: 240)

Bringing together the abstract and physical ways of characterizing autonomy, we can state in general terms that an autonomous system is a thermodynamically open system with operational closure which actively generates and sustains its identity under precarious conditions. The paradigm case of an autonomous system is the living cell. The constituent processes in this case are chemical; their recursive interdependence takes the form of a self-producing, metabolic network that also produces its own semi-permeable membrane; and this network constitutes the system as a unity in the biochemical domain and determines a domain of interactions with the world. The notion of precarious conditions comes from Ezequiel di Paolo (2012). The underlying processes in this case are chemical, with recursive interdependence in the form of metabolic, self-producing methods. This, being responsible for producing its own semi-permeable membrane, constitutes the system as a unity in the biochemical domain, which ultimately determines a domain of interactions with the surrounding world.

Following Maturana and Varela (1980) and Varela et al. (1974), autopoiesis is defined as a kind of self-producing autonomy at the molecular level. The nervous system, sensorimotor system, the multicellular bodies of metazoan organisms, and the immune system are examples of autonomous systems (Thompson and Stapleton 2008).

Another proposition, proposed by Clark and Chalmers (1998) as an extended mind thesis is that the environment constitutes part of the mind when it is coupled to the brain in the right way. But both viewpoints stress the crucial contributions regarding the body and the environment taking part in the process of cognition. Recent discussions on this model have emphasized how these views also differ in significant ways (Clark 2008; Wheeler 2008). According to them, this can be introduced as an operational closure of the nervous system rather than considering it as a closed system.

As for preliminary concepts, we know that CNS is that portion of the vertebrate nervous system which is composed of the brain and spinal cord. Together with the peripheral nervous system (PNS), the other major portion of the nervous system, i.e., the CNS, coordinates the body's interaction with the environment. This area can be described as neither uniform nor composed of similar modules but rather consists of a mosaic of highly interconnected and different regions. Accordingly, knowledge about the brain regions, i.e., how these regions are connected functionally, is crucial to gain the proper knowledge about the behavior, perception and cognition in different parts of the brain. Physically, this bodily system is contained within the dorsal cavity, with the brain in the cranial sub-cavity (the skull) and the spinal cord in the spinal cavity (within the vertebral column). Now, in thinking of the brain as a "closed system", we do so with respect to the exchange of the intrinsic recurrent interactions of neuronal circuits with the environment but which is open to the exchange of matter (at the molecular level), i.e., information or energy. It is to be mentioned that, in dealing with this kind of problems, functional connectivity methods are successfully applied in estimating similarities between activities recorded in different regions of the brain. Thus, within the approach of functional geometry, the brain is viewed as embedding intrinsic degrees of freedom that are open to the exchange of sensory input which modulates the on-going intrinsic "closed activity". This is very much like changing the direction of the spin of an elementary particle when the particle is placed in an external magnetic field.

The CNS was also considered as a closed system by Hamburger (1963). However, he also indicated that the basic organization of the CNS is responsible for the generation of internal states (Hamburger 1963) rather than in the form of "homuncula vernacles". In the Hamburger era, due to the strong influence of cybernetics, those working in this line of thought conceived the CNS to be a system connected primarily to information processing. It is now a quite established fact that the brain's various functional networks spontaneously fluctuate with activities even "at rest", correlated to the different intensities of their activity levels. The multiple "temporal functional modes" have already been identified. These are subdivided into corresponding default-mode networks (but with the regions anti-correlated with it), which again, can be divided into several functionally distinct, spatially over-lapping, networks, each having its own patterns of correlations and anti-correlations. These functionally distinct modes of spontaneous brain activities are, in general, quite different from the resting-state networks previously reported and may have greater biological interpretability. The brain functions continuously through a remarkably high level of intrinsic activities and also, with non-stationary dynamics. But, interestingly, it reveals highly structured patterns across several spatial scales, starting from fine-grained functional architecture in sensory cortices to large-scale networks. But until now, the nature of this particular kind of activity is poorly understood. However, the response to a sensory input could possibly be availed from appropriate motor output which is properly computed. Many of recent research findings suggest the existence of motor activities even before the maturation of the sensory systems. Also, the idea that **the senses only influence behavior but do not dictate it** helped to launch the conception of the CNS as an **autonomous system.**

The nervous system evolved in such a manner that multi-cellular creatures moving in a purposeful fashion, e.g., non-randomly, have a 'clear selective advantage'. To accomplish such 'intelligent movement', the nervous system evolved a set of strategic and tactical rules, i.e., prediction which is nothing but the ability to anticipate the outcome of a given action. This is based on the incoming sensory stimuli and previously learned experiences or inherited instincts. Thus, it can be opined that the ability to predict the outcome of future events is the most universal and significant of all global brain functions, though with relevant controversial points yet to be argued. In this regard, Hommel et al. (2001) considered well-ordered actions as the selective advantage of intelligence. He argued that perception is, fundamentally, a part of action having interrelationships between perception and action. As the regularities in coding become established, this idea conceptualizes the understanding of neural learning, which can occur either in perception or motor control. Any other well-correlated system that can easily be linked directly with it will eventually be capable to monitor their activities. It is generally held that these activities explain fully the separation of the sensory and motor cortical maps. Interestingly, exceptions are found in the case of most primitive mammals. There, in certain circumstances, this separation provides greater opportunity for the intervention of between-process action control (Livesey 1986). Sometimes, perception and action are fused even more tightly than what the feature-based coding implies. Tanji and Shima (1994) studied trained monkeys performing sequences of actions, either from memory or by following cues. According to their observations:

> It was certainly not by accident that work on attention had a fresh start with the advent of the information-processing approach (for historical reviews, see Neumann 1996; Van der Heijden 1992). Attentional mechanisms account for the various limitations observed in human performance. Thus, they are assumed to enhance or to inhibit the activity in the various streams and at various stages in the flow of information. The details of attentional theories are assumed to depend on the details of the supposed cognitive architecture (Tanji and Shima 1994).

Another problem lies in the level of arbitration between behaviors: the problem, mentioned earlier, is how to choose appropriate contexts in which to act, or to attend to the relevant behavioral processes. It is a well-established fact that arbitration is necessary in a parallel, distributed model of intelligence, because the finite resources (e.g., hands and eyes) possessed by an agent, must be shared among various possible behaviors (e.g., feeding). Hence, arbitration must account for both the cases of the activation level, present in the various 'input' cortical channels, and also for the previous experiences, i.e., both of the current or related action-selection contexts.

The next immediate problem arises regarding the mechanism of prediction, which can be termed as a ubiquitous one, appearing in the brain's control of movement. Diedrichen et al. (2002) focused on the problem of coordination where multiple effectors (joints, limbs and/or muscles) play a vital role. Their theory and its more recent extension, i.e., optimal feedback control theory, provide valuable insights into the flexible and task-dependent control of movements. This model also makes quantitative predictions concerning the distribution of work across multiple effectors. In addition, it predicts variation in feedback control, ultimately demanded for the changes needed to be worked out, that are required by the tasks involved. Working out in detail the correlation structure between different effectors, e.g., muscles, limbs or joints, they suggested that movements are made for achieving goals whereas effectors coordinate the role of task-relevant states associated with the body and environment.

Next, let us take the example mentioned in their works (Diedrichsen et al. 2009): at the time of reaching to press an elevator button, the task-relevant state can be considered as the position of the index finger, whereas the goal of the reaching is to bring the fingertip to the button. Now, a fundamental problem of coordination turns out to be that the number of effectors involved (as in this example, there will be 10 possible degrees of freedom of movement between shoulder and index finger, and the number of muscles will be more than 40, for performing those movements) exceeds the dimensionality of the task requirements (three spatial dimensions). It is possible to employ many different ways to achieve the goal of movement. But even after possessing this inherent redundancy, a large body of experimental data obtained from them indicates that the motor system consistently uses a narrow set of solutions. Then, one of the central issues in the research on coordination turns out to be the problem of out how and why the brain selects particular movements, given the large set of possibilities.

Consider the simple act of reaching for a carton of milk in the refrigerator. Without giving much thought to our action, we must predict the carton's weight and the compensatory balance that we must apply for a successfully smooth trajectory of the carton from the refrigerator. Indeed, before even reaching, we have made a ballpark prediction about what we are going to be involved with. And surprisingly, we often hit the carton against the shelf above because it is actually lighter than we had unconsciously predicted. Furthermore, we can assume that the prediction in movement must be unique (i.e., the CNS can only operate in a probability space having only a one-dimensional axis), and so, predictive functions must be collapsed finally to a one-dimensional vector to be the basis for motor prediction at any given moment. Concerning the proper mathematical format needed to understand brain–decision theory, the possibility of using Bayesian statistical methodology has raised much interest (Körding 2007) The possibility of using the concept of functional geometry associated with neuronal network properties to understand the predictive properties of the brain was initially proposed in the late 1970s and early 1980s by Pellionisz and Llinás (1979, 1985). Then the spatio-temporal nature, together with the metric property, has been addressed more recently from a dynamic perspective, rather than as a linear connectivity matrix transformation, under the term **dynamic geometry**, where the metric is considered to be statistical in nature (Roy and Llinas 2008). The functional basis for such a kind of statistical metric view of internal functional space arose from the experimental finding that inferior olive neurons are endowed with weakly-chaotic, intrinsic oscillatory properties (Makarenko and Llinas 1998). Henceforth, as with the initial network tensor approach, dynamic geometry transforms a dynamic covariant vector (associated with sensory inputs) into a contravariant dynamic vector (associated with motor outputs). From this perspective, the distribution associated with the statistical metric can predict the movements uniquely. As a result and in addition, the application of Bayesian methodology (Körding 2007) to decision making becomes an attractive avenue. In particular, as a well-known fact, one of the most significant issues in the Bayesian approach concerns the genesis of priors (prior belief or prior knowledge) in neuronal circuits.

3.1 Prediction and Movement

Since the ability to predict has evolved in tandem with increasingly complex movement strategies, we may address movement control in order to understand prediction. Let us return to the above mentioned example, i.e., removing a carton of milk from a refrigerator. The appropriate pattern of contractions must be specified for an extension/grasping sequence to be properly executed. Each muscle provides a direction of pull (a vector) composed of individual muscle fibers operating in an pre-established manner, based on their common innervations by given motor neurons. These contractile ensembles are known as motor units (a single motor neuron innervates tens to hundreds of individual muscle fibers). A given muscle may be composed of hundreds of such individual motor units. The number of muscles multiplied by the number of motor units in each muscle may then be viewed as the total number of degrees of freedom for any given movement.

A movement such as reaching into the refrigerator might appear as a simple one, but from a functional perspective, even a simple movement often engages most of the body's muscles, resulting in an astronomical number of possible simultaneous and/or sequential muscle contractions and degrees of freedom. In addition, the arm may be brought towards the milk carton from any number of initial positions and postures (maybe your back hurts, so you reach from an unusual stance). All of this potential complexity exists before the load is actually placed on your body; you have yet to pick up the carton and can only guess its weight during your initial reaching motion. So, this simple movement is not simple at all, when we break it down and try to understand how the brain handles all of this. However, the dimensionality of the problem of motor control does not derive solely from the number of muscles involved, the various degrees of pull force and angle and so forth. The real dimensionality of the problem stems from the complicated inter-action between the possible directions of muscular pull and their temporal sequence of activation. Much of motor control occurs in real time, 'online', as it were. Our movements, from start to finish, are seldom under stimulus-free conditions.

Next, consider the following scenario: holding a cup of coffee but simultane-ously riding your bicycle. The combination of muscle contraction and relaxation at any given moment is often determined as a movement sequence and executed in response to teleceptive stimuli (hearing and vision), kinesthetic feedback (your feeling of the bicycle and of the cup of coffee) and your intentions (drink your coffee and get to the your destination as soon as possible). Optimal controller is generally assumed to be one producing the smoothest possible movements. To minimize the accelerative transients can only be achieved at the expense of pro-ducing jerkiness in movement. This requires almost accurately continuous moni-toring, at a sampling rate on the order of milliseconds or faster (e.g., your auditory system), and also "feed forward and feedback" influences on the selected activation sequences. Yet, though this may appear surprising, quantitative evaluation is computationally possible, for the brain to control our movement in such a con-tinuous manner. From the heuristic description above, and given that there are 50 or so key muscles in the hand, arm and shoulder, so as to reach the milk carton, over 10^{15} combinations of muscle contractions are possible—a staggering number, to say the least. If during every millisecond of this reaching/grasping sequence, the single best of the 10^{15} combinations is chosen after an evaluation of all of the possibilities, then 10^{18} decisions would have to be made per second. This would require of the brain, if it were to a computer to possess a one-EHz (one-million GHz) processor which would be necessary when choosing the correct muscle combinations to successfully execute this, seemingly, relatively simple, reaching/grasping sequence. In reality, the dimensionality of the problem of motor control increases many orders of magnitude when one considers that there is a bare minimum of 100 motor units for every muscle, and that each muscular pull may, and most likely will, involve various sets of motor neurons. The brain seems to have evolved to deal with the control of movement in the way described above, if one considers that approximately only 10^{12} neurons are present in the entire brain. So, the continuous control of movement demands an extremely high computational overhead. This will be valid only if the brain is controlling the movement by regulating the activity of every muscle discretely, in parallel, or by choosing and implementing combinations of muscles, when making complicated movements frequently. To delve further into this issue, we must ask the following questions:

- How might the dimensionality problem of motor control be reduced without significantly degrading the quality of movement sequences?
- Which aspects of brain function may provide clues to solve this problem?

To achieve a common goal, a common understanding should be reached on the central problem in motor control which is based, primarily, on the understanding about the role played by many biomechanical degrees of freedom having coordination among them. What is a special and a puzzling aspect with respect to coordination is the behavioral goals that can be achieved with consistent reliability. Not only that, but so long as the repetition of the movements is concerned, the coordination and corresponding control can rarely be reproduced in detail. Existing theoretical frameworks emphasize either of two, i.e., the goal achievement or the richness in motor variability, but fail to reconcile them.

Emanuel and Jordon (2002) observed a range of motor tasks obtained from their experimental results. They supported their findings by proposing stochastic optimal feedback control as an alternative approach. Their theory permits the variability in redundant (task-irrelevant) dimensions with which it is possible to achieve an optimal strategy in the face of uncertainty. They corrected only those deviations that are found to interfere with task goals. Thus, their strategy does not enforce any desired trajectory, but ensures the feedback in a much more intelligent way. Altogether, this framework established the solutions for many difficult tasks associated with their theory, i.e., how to solve many modes of criteria, e.g., (i) the task-constrained variability, (ii) goal-directed corrections, (iii) motor synergies, and (iv) controlled parameters, simplifying the rules and discrete coordination modes when all of them emerge naturally. However, there are other schools of thought, for example, Arbib (2007) that focus on the evolution of action capabilities instead of considering the evolution of the brain to have such capabilities. They proposed evolution for yielding a language-ready brain. In this regard, several examples of simple imitations have been proffered, for example, as seen in great apes and their complex, goal—directed imitations. This has also been noted in the case of humans. The resultant suggestions point to the possibility of considering the presence of action sequencing there, as a background for the analysis of modelling strategies. A relatively straightforward approach in reducing the dimensionality of motor control would be to decrease the temporal resolution of the controlling system, i.e., to remove it from the burden of being continuously online and processing. This can be accomplished by breaking up the time line of the motor task into a series of smaller units over which the controller must operate. Control would be discontinuous in time, and thus the operations of such a system would occur at discrete intervals of a '∂t. We must consider an important consequence of this approach, namely, that the movements controlled by this type of pulsatile system would not be executed continuously, in the sense of demonstrating obligatorily smooth kinematics, but rather would be executed in a discontinuous fashion with a series of muscle twitches that are linked together. Motor physiologists have known this fact for over a century: *"movements are not executed continuously, but are discontinuous in nature"*. Indeed, as early as 1886, E.A. Schafer indicated that human movement was discontinuous:

The voluntary of a muscular contraction... invariably shows, both at the commencement of the contraction and during its continuance, a series of undulations that succeed one after another with almost exact regularity, and can... only to be interpreted to indicate the rhythm of the muscular response to the voluntary stimuli which provokes the contraction.... The undulations... are plainly visible and are sufficiently regular in size and succession to leave no doubt in the mind of any person who has seen a graphic record of muscular titanic contraction produced by exciting the nerve about 10 times in the second, that the curve... is that of a similar contraction. (On the Rhythm of Muscular Response to Volitional Impulses in Man, Schafer E.A., (1886); The J. Phsiology, DOI:1113/jpysiol.1886.sp000210).

3.2 How Does the Brain Predict?

Enhancement of brain function entails controlling neuronal function. For the nervous system to predict, it must perform a rapid comparison of the sensory-referred properties of the external world with a separate internal sensorimotor representation of those properties. For the prediction to be successfully realized, the nervous system must then transform it into, or use, this pre-motor solution in finely timed and executed movements. Once a pattern of neuronal activity acquires internal significance (sensory content gains and internal context), the brain generates a strategy of what to do next, i.e., another pattern of neural activity takes place. This strategy can be considered as an internal representation of what is to come, a prediction imperative, in order to become actualized in the external world.

3.2.1 Motor Binding in Time and the Centralization on Prediction

3.2.1.1 Pulsatile Motor Control System

This fact has already been established that the brain can lower the computational work load for controlling movement (motor output) by sending motor control signals in a non-continuous and pulsatile fashion. Following Llinas and Roy's (2008) proposal, the idea that synchronizing sensory information with pulsing motor control signals brings to mind more poetic notions of rhythmicity and the way that the idea "yogis" use their breath to enhance and unify their outer and inner world experience. Our brains have enormously complex computational tasks to perform and automatically enhancing brain function entails controlling neuronal function. Thus, the brain may have just another strategy to use less labor and achieve more fruitful results: it has evolved to simplify the enormous computational load associated with moving and coordinating the body. This computational strategy has the added benefit of making it easier to bind and synchronize motor-movement signals with a constant flow of sensory input. As Moshe Bar (2009) put it:

> As is evident from the collection of articles presented in the brain might be similarly flexible and 'restless' by default. This restlessness does not reflect random activity that is there merely for the sake of remaining active, but, instead, it reflects the ongoing generation of predictions, which relies on memory and enhances our interaction with and adjustment to the demanding environment

Neural circuits in the body and brain are inherently good at learning and storing information which makes them very efficient in predicting what to do with incoming sensory inputs. This may just be an efficient strategy that the brain has evolved to simplify the enormous computational load associated with moving and coordinating the body. First, from a control system and, next, a pulsatile input into motor neurons from a control system, as opposed to a command system, may prepare a population of independent motor neurons for descending control by uniformly biasing these motor neurons into their linear range of responsivity (Greene 1982). To clarify, a pulsatile control input would serve to 'linearize' a population of highly nonlinear and independent neuronal elements to ensure a uniform population response to a control signal. The motor neurons that need to be recruited for a given movement may be, and often are, separated by many spinal levels; this mechanism may be serving as a cueing function to synchronize motor neuronal activation. Secondly, a pulsatile control system might allow for brief periods of movement acceleration that would provide an inertial break mechanism to overcome frictional forces and the viscosity of muscles (Goodman and Kelso 1983). For example, when we rock a snowbound car, this movement helps to extract it. The third point lies in the fact that a periodic control system may enable input and output to be bound in time. In other words, this type of control system might enhance the ability of sensory inputs and descending motor command/controls to be integrated within the functioning motor apparatus as a whole. We now understand that operating continuously online, which we might have thought of as the only way the brain could bring about smoothly executed movements, is simply not possible physiologically. Instead, the brain has relegated the rallying of the motor troops to the control of a pulsatile, discontinuous signal, which is reflected in the musculoskeletal system as a physiological tremor. Other than just saving the brain from being computationally overwhelmed, a pulsatile control input also serves to bring the neurons, muscles or limbs closer to the threshold for some action, be it firing, integration or movement. The possible risks of operating discontinuously in time are beautifully minimized by the synchronizing effect. This pulsatile signal has the independent elements of the motor apparatus at all levels. Let us remember the words of Berstein (1967): "a mutual synchronization through rhythm is doubtless necessary for the motor apparatus as a whole". His realization was that the factors taking part in the system are mechanical, subject to gravitational and inertial forces, as well as to reactive forces created by the movements of links in between the biokinematic chains. Consequently, there is no simple 'one to one' relationship between a physical stimulus and a psychological percept. The impulses and their outcome in movement must be indeterminate, very similar to the problem posed by the perceptual theorists. The different "contextual conditions" of a movement depend both on environmental changes as well as on the dynamic state of the component parts, for example, body parts, i.e., mass. Once implemented, this mass gathers momentum, develops kinetic energy, ultimately providing forces of action with eventual effects on other parts or segments. We are aware that the muscles are often used in combinations. Fixed or hard-wired synergies are not the only rule, and muscle combinations clearly change dynamically, which they must do during the execution of a complex movement.

Again, it is an already observed fact that, for feeding behavior, it is required to have intricately coordinated activation among the muscles of the jaw, tongue and face. But, the neural anatomical substrates that underlie such coordination are still to be clarified (Stanek et al. 2014). Two groups of premotor neurons have been identified as responsible for the distribution in the regions. These regions are implicated in rhythmogenesis, descending motor control and sensory feedback. Also, several premotor connection configurations are found, not only ideally suited for coordinating bilaterally symmetric jaw movements but also for enabling the co-activation of specific jaw, tongue and facial muscles. All these facts indicate that shared premotor neurons form specific multi-target connections with selected motor neurons which can be concluded to be a simple, but general, solution to the problem of orofacial coordination. Thus, it can be stated that behaviors are executed through the coordinated activity of multiple groups of motor neurons and their muscle targets. Coordination of jaw and tongue muscles during feeding behaviors represents one of the most intricate mechanisms of the motor system. These characteristic facts have been observed in a wide range of animals, including humans.

In summary, a set of pre-emptor connection configurations has been discovered, well-suited to enable multi-muscle coordination and bilateral symmetry observed in feeding behaviors. Shared pre-emptor neurons, simultaneously providing inputs to multiple groups of motor neurons, innervating specific muscles, may be a common circuit mechanism for motor coordination. At this point, now the following question arises: "*if muscle collectives are the unit to be controlled, as opposed to individual muscles, then what does this ask of the central process underlying movement control?*" The answer is that, whenever a complex movement proceeds, it demands the control system to be able to reconfigure itself dynamically. These collectives are cast temporarily, quickly dissolved and rearranged as and when required. If the CNS is considered to have many possible solutions for a given motor task, then it automatically follows that any given functional synergy organized by the brain must be a fleeting, dissipative construct. Furthermore, such constructs may not be easily recognized in behavior as an invariant pattern of muscle activation, such as those we recognize in many overt, stereotypical reflexes. An 'over complete' system of muscle collectives would ensure a degree of versatility and flexibility in the choices that the control system could make. If we think all of the different ways we can reach for the milk carton, the idea of over completeness is clear. If the motor control system may select from an over complete pool of similar functional synergies, any number of which gets the job done reasonably well, then this would certainly lower the burden for the control system. As a final result, this phenomena would ease the demand for precision, i.e., for having to make the right choice every time.

3.3 How Can a Neuronal Circuit Predict?

In one of the early attempts to understand prediction, Pellionisz and Llinas (1979) concluded that the brain predicts "by taking advantage of the differences in electrical behaviour between the given nerve cells".

The ways in which neurons receive and process information determine how the brain functions. For example, in order to make a decision, the ways in which evidence is accumulated for various, competing, alternative neurons, eventually allow an animal to make one choice over another. These accumulating neurons are termed "neural integrators". This kind of integrator can be called short-term memory systems, which store a running total of the inputs they receive. While easiest to understand in terms of decision making, neural integrators show up in a variety of other brain processes, too. For example, when signals that contain information about the velocity of movements are integrated, they become signals for navigation, i.e., motor controls that contain information about body position. The accumulation and storage of information can be context-specific in any of these cases, i.e., decision making or motor control. This leads to the condition that different information is accumulated depending on the conditions under which a task is performed: in other words, the task's context.

Context specific accumulation and storage of information in a well-studied system of the brain has been investigated by an experimental group (Goldman 2000).

For example, let us consider the case of the oculomotor system which controls eye movements. This system contains a neural integrator circuit that converts eye velocity signals into eye position signals. The nature of the conversion has been noticed as context-specific, i.e., the circuit maintains distinct patterns of neural activity depending on whether a given eye position has been reached through a sudden, rapid eye movement or a slower, tracking eye movement. Their works promise to reveal ubiquitous mechanisms in which neurons have been found accumulating information and storing it in short-term memory. Such information is applicable not only to motor systems, but also to higher-level cognitive processes such as decision making.

Rosenblatt (2000), from his experimental findings, introduced a novel idea about using "utility" fusion. An alternative to both sensor fusion and command fusion, a new method of action selection is introduced where "utility fusion" is applied via fusion. But, it is a well-known fact that the distributed asynchronous behaviors indicate the utility of various possible states, including their associated uncertainty. In such a case, for a centralized arbiter, it is possible to combine these utilities and probabilities in order to determine the optimal action based on the maximization of expected utility. This map, based and constructed on all these factors, enables the system to be controlled, modelled as well as compensated for. Experimental results verified that this approach improves performance by which a solution to the shortcomings of command fusion and sensor fusion systems can be reached. Both the distributed and asynchronous behaviors, instead of voting for action, indicate the utility of various possible world states together with their probabilities, based on domain-specific knowledge. Then, various candidate actions can be evaluated by the arbiter, who employs those system models, dictating the proper actions to be taken without violating kinematic and dynamic constraints. This technique provides us with an expectation to gain greater stability but without time-dependent utility. Using this kind of representation, an arbiter is capable of synchronizing, and maintaining effectively, a consistent interpretation of the votes which are received from asynchronous behaviors. The utility-fusion technique combines not only the advantages of sensor fusion and command fusion, but also provides coherent, optimal reasoning in a distributed, asynchronous system with well-defined semantics of votes and uncertainty, producing experimental results with better control and measurability and avoiding many of their drawbacks.

Again, in the case of some neurons that are highly sensitive to stimuli while others are less so, a ∂t 'look ahead' function might be implemented by neuronal circuits through a process analogous to a mathematical function, known as the Taylor series expansion. But the question arises here: "*what is at the heart of such central circuits that could provide the intrinsic drive to generate organized movement if it is not a reflex/sensory input*"? The self-referential approach to the organization of motor control has been given a shot in the arm by the discovery of intrinsic neuronal oscillations and the specific ionic currents necessary for their generation (Llinas 1988). Indeed, oscillatory neuronal behavior is associated with the generation of an overt, rhythmic activity at 8–12 Hz. Thus, the periodic activity seen in movement is a reflection of a motor control system operating in a pulsatile, discontinuous-in-time fashion. A control system that synchronizes motor control signals temporally so that movement is executed in an organized, expedient manner must be centrally located. The spinal cord is more than capable of sustaining a rhythmic movement, but it does not have the wherewithal to organize and generate a complex predictive movement. From their findings, Wessberg and Kakuda (1995) reported another important piece of information, namely that constant velocity $(4° \text{ s}^{-1}$ to $62°\text{s}^{-1})$ finger movements are produced by the pulsatile discontinuities, which are present in the descending motor command. These findings demonstrate the fact that centrally derived rhythms modulate the firing of motor units, which are of physiological significance (Farmer 2004). The Göteborg group, after observing human subjects for a substantial period, found discontinuities in slow finger movements. However, over a century ago, this kind of phenomenon was observed and a small amount of discontinuity was noted to appear in the finger position trace, even after hard efforts made by the subjects for producing a completely smooth, slow movement. Before the landmark works of Valbo and Wessberg (1993), physical tremors were generally assumed to be an imposition on the movement instead of considering these $\sim 8\text{--}10$ Hz discontinuities as representing a physiological tremor. Valbo and Wessberg (1993) discussed the difference between the mechanical characteristics of movement-related discontinuities and those of tremor; because, during very slow movements, the peaks of acceleration and deceleration are observed to be asymmetric, interspersed and having large amplitude. Not only that, but instead of there being a small sinusoidal oscillation around a neutral point, periods of movement are noted, which have 8–10 Hz intervals, and also, occur close to the peak acceleration. After the analysis of the EMG during movement, a burst was revealed when the muscle acts as "agonist" and it comes close again to peak deceleration when the muscle acts as "antagonist". It is interesting to note that they observed, furthermore, an alternating pattern of EMG indicating that the minute movement steps are achieved through pulsatile agonist-antagonist activities. Pursuing further their experiments furthermore, the same group also observed that the firing of a different group's reflex afferents were modulated by the 8–10-Hz discontinuities. This suggests that, actually, in generating the discontinuities (Wessberg and Valbo 1996), there exists a resonance in between the spinal reflex circuitry and the signal (Wessberg and Valbo 1995). However, there is no evidence to establish the primary role of the spinal and supraspinal stretch reflex when temporal relationships between EMG, acceleration and afferent firing are studied. Though the various lines of evidence support a centrally imposed 8–10-Hz signal, a possibility still remains that, during movement, discontinuities might occur due to the presence of sub-tetanic motor-unit firing at 8–10 Hz. In order to justify this hypothesis, Wessberg and Kakuda (1999) calculated directly the coherence between simultaneous recordings of single motor-unit activity and the acceleration signal during different velocities of movement.

According to them, their results show 8–10-Hz coherence between motor unit firing and the acceleration. This indicates that the two signals are co-modulated. The frequency of the coherence is found to be the same as that of the movement discontinuities. But, it is important to mention that it is quite different from the motor-unit firing rate where the temporal modulation pattern occurs, rather than the motor-unit firing rate. This is instrumental in the production of movement discontinuity. While to date pairs of motor units have not been studied with this protocol, it seems that during movement, an 8–10-Hz signal is imposed on the motoneuron pool, producing 8–10 Hz co-modulation of individual motor units that may be firing at different rates. This co-modulation is then expressed as movement discontinuity. In the context of the previous data, describing out-of-phase co-modulation of agonist and antagonist motor neurons, an interesting model emerges in which slow movements are produced through a discontinuous (8–10 Hz) command, generating small pulses of acceleration and braking. Increasing movement velocity is associated with larger amplitude discontinuities, suggesting that the amplitude of the pulses sets the velocity of movement. As previously highlighted by the group, there is an interesting parallel with ballistic movements in which the triphasic 'agonist-antagonist-agonist' EMG bursts arrive in a similar temporal pattern to that detected for 'constant'-velocity finger movements.

The idea that motor unit firing may be modulated by different frequencies above, below or at the mean motor unit firing rate, has many interesting implications for motor control. During isometric contraction, motor units and acceleration may be modulated by frequencies in the ranges 1–12 and 16–32 Hz, whereas the higher frequency range is coherent with signals recorded from above the primary motor cortex. During a maintained contraction, 20-Hz supraspinal oscillations co-modulate motor unit activities both within hand muscles and between closely related agonist muscles. But in the moving state, it is dominated by the 8–10-Hz signal (Farmer 1998). Another important finding from the experimental results of Wessberg and Kakuda (2004) and Wessberg (2004), reported that constant velocity (4° to 62°) finger movements are produced by pulsatile discontinuities in the descending motor command. These findings are of great physical importance as this provides further evidence that the firing of motor units in humans is modulated by a centrally derived rhythm.

Erimaki and Christakos (2008) studied firing muscle force oscillations having 6–10 Hz motor units (MU), where they emphasized their neural substrate as well as the relationship with rhythmic motor control. They experimented with 24 human subjects and performed their observation of finger muscles during 121 contractions. From their experimental results, they came to the following conclusion:

> Coherent 6- to 10-Hz components of MU discharges coexist with carrier components and coherent modulation components underlying the voluntary force variations... 6- to 10-Hz synchrony has the frequency of the tremor synchrony in steady contractions, widespread, in-phase, strength ranges from very small to very large (MU/MU coherence > 0.50) among contractions; ... related to the contraction parameters, in accord with notion of a distinct 6- to 10-Hz synaptic input to the MUs. Unlike the coherent MU modulations and the voluntary force variations, the in-phase 6- to 10-Hz MU components... suppressed or even eliminated ...respective force component ... drastically reduced (Erimaki and Cristakos 2008).

Their findings support the widely assumed supraspinal origin of the motor unit (MU) modulations, an essential role for muscle spindle feedback during the generation of the 6–10-Hz synaptic input providing important information about the study of 6–10-Hz rhythm generations. This subserves the postulated rhythmical control, manifested as force, having movement components. They proposed the participation of spinal stretch reflex loops having an oscillating nature in the rhythm generation, possibly, during the interaction with supraspinal oscillators. But little is known about the anatomical substrates of these signals, although, it is tempting, in the light of evidence from animal experiments, to postulate that 20-Hz oscillations are derived from synchronous activities, present in the primary motor cortex, whereas, cerebellar and sensory pathways (Farmer 1998) observed in animals and man during reaching and other more proximal movements. But, it is not yet known whether the movement related 8–10-Hz co-modulation of motor units and acceleration is a general phenomenon. Many more factors, e.g., (i) the relationships and interactions between 8–10-Hz-movement-related frequencies, (ii) 8–12-Hz physiological tremors and (iii) 20-Hz-frequencies associated with position holding remain yet to be elucidated. However, the results of Wessberg and Kakuda (1999) and other recent findings suggest that all these frequencies mentioned, i.e., 8–10 Hz, 20 Hz and other rhythms form an important part of the syntax of movement and posture. Stanek et al. (2014), using a modified monosynaptic rabies virus based on trans-synaptic tracing strategy, investigated in detail whether the pre-emptor circuitry of jaw and tongue motor neurons contain elements for coordination. They (Stanek et al. 2014) explained using:

> a modified monosynaptic rabies virus-based transsynaptic tracing strategy, we systemati-cally mapped premotor neurons for the jaw-closing masseter muscle and the tongue-protruding genioglossus muscle. The maps revealed that the two groups of premotor neurons are distributed in regions implicated in rhythmogenesis, descending motor control, and sensory feedback. Importantly, we discovered several premotor connection configu-rations that are ideally suited for coordinating bilaterally symmetric jaw movements, and for enabling co-activation of specific jaw, tongue, and facial muscles. Our findings suggest that shared premotor neurons that form specific multi-target connections with selected motor neurons are a simple and general solution to the problem of orofacial coordination. DOI:10.7554/eLife.02511.001.

These results establish the fact that the two groups of premotor neurons are distributed in regions implicated in rhythmogenesis, descending motor control and sensory feedback. Importantly, it has been now a quite established fact from the discovery that these kinds of several pre-emptor connection configurations are ideally suited for coordinating bilaterally symmetric jaw movements. Their findings suggest that shared premotor neurons, forming specific multi-target connections with selected motor neurons, are a simple and general solution to the problem of orofacial coordination.

3.4 Dynamic Geometry and Bayesian Approach to Decision Theory

If, as stated above, there are 50 or so key muscles in the hand, arm and shoulder that one uses to reach for the milk carton, over 10^{15} combinations of muscle contrac-tions are possible. The choice of appropriate motor command is primarily a decision process. However, in addition to muscular combinatorial issues, noise in the sen-sory inputs generates uncertainty in the hand's true location. This type of uncer-tainty contains the problem of estimating the state of the external world. Bayesian statistics provides a systematic way of approaching such a problem in the presence of uncertainty (Cox 1946; Jaynes 1986; Freedman 1995; Mackay 2003). In this kind of statistics, prior knowledge about the system is assumed. The fundamental question regarding the CNS function is the origin of such prior knowledge, and the answer is simple: the morpho-functional brain network is initially inherited and then honed by use. In fact, the origin of a Bayesian framework can be traced to Helmholtz's (1867) idea of perception. It was Helmholtz who realized that retinal images are ambiguous and prior knowledge is necessary to account for visual perception. Recently, a genetic component has been thought to play a significant role in the development of human depth perception (Yonas 2003). And so the next question is: *How is prior information encoded in the CNS?*

A central question concerning present-day neuroscience is that of understanding the rules for the embedding of 'universals' into intrinsic functional space. Some suggestions have been offered concerning the nature of this CNS internal space which we (Roy and Llinás 2008), address. The arguments go as follows: if functional space is endowed with stochastic metric-tensor properties, then there will be a dynamic correspondence between the events in the external world and their specification in the internal space. We shall call this dynamic geometry because the minimal perceptual time resolution of the brain (10–15 ms), associated with 40-Hz oscillations of neurons and their network dynamics, is considered to be responsible for recognizing external events and generating the concept of simultaneity (Joliot et al. 1994).

Essentially, this dynamic geometry helps us understand the nature of sensory–motor transformation. The stochastic metric tensor, in this geometry, can be written as a function of four space–time coordinates (i.e.. three spaces and one Minkowski space) and a fifth dimension (Roy and Llinás 2012). This fifth dimension is a probability space as well as a metric space. This extra dimension is an embedded degree of freedom and random in nature. In fact, the dynamic geometry framework makes it possible to distinguish one individual from another. The uniqueness of individuals must operate in detail, not in principle, i.e., general solutions must operate on the basis of a generalized solution field. Considering the stochastic nature of specific brain function among individuals, they will have both anatomical and functional variances, similar to the differences in their facial characteristics. The variances must be such that they modulate, rather than destroy, the general solution. However, the variances may be so extreme that they negate the function or may even be lethal. Within this framework, social variance enriches the system, whereas some variances augment some properties and diminish others, so that the individuality is quite an important factor. Thus, the geometric structure is embedded in the CNS from birth, and functional states of neurons are associated with this geometric structure. In this framework, the covariant-to-contravariant transformation occurs with a certain probability distribution.

They explained the "physiological tremor" as the reflection of T-type calcium channel kinetics in inferior olive neurons (Llinás et al. 2004), as a result of which the fluctuating nature of the geometry of the internal space occurs. Here, the input–output feedback is not simply sequential but related to the stochastic process in equilibrium whose distribution is associated with the distribution of the stochastic metric tensor. This gives rise to an optimal (and organized) motor movement. As stated earlier, Bayesian methodology helps to explain the motor and predictive ability of brains. This is basically an issue in statistical inference, which can also be viewed as information processing relating input and output information. Also, we know that the statistical inference, studied in various topics in statistics, physics, biology and econometrics, is a form of information processing where a black box endowed with prior information (morpho-functional properties) receives input information and sends outputs in the form of parameter estimates and predictive distributions. In fact, Helmholtz's idea of perception (1867) as 'unconscious inference' is considered as the origin of Bayes' framework. Bayes' theorem, widely studied as a model to transform a priori information into a future inference distribution, is basically explained in the following manner:

Given a prior distribution, say a Gaussian distribution $p(x)$ and noisy observation o (data), which leads to a $p(o|x)$, then the p inference distribution $p(x|o)$ is given by:

$$P(x|o) = p(o|x)p(x)/p(o)$$

Körding (2007) developed a Bayesian approach, which leads to a better estimate of the possible outcomes than other estimates using sensory input. Recently, Bhattacharya et al. (2007) developed a Bayesian methodology to fit the data observed in an underlying process in equilibrium. This methodology can be described as:

$$f(\text{data process, parameters}) \times f(\text{process}|\ \text{parameters}) \times f(\text{parameters}).$$

Which, when expressed in a formal language is:

$$f(Y|X, \theta)f(X|\theta)f(\theta);$$

where, Y is the (set of) observed data; θ is a set of model parameters; and X is an unobserved (latent) stationary stochastic process. This stochastic process is induced by the first-order transition model:

$$f(X(t+1)|X(t), \theta)$$

where, $X(t)$ denotes the state of the process at time t. It is to be noted that given θ, the transition model

$$f(X(t+1)|X(t), \theta)$$

is known, but the distribution of the stochastic process in equilibrium, i.e., $f(X|\theta)$ is generally unknown. The crucial feature of the above type of model is that, given θ, the transition model $f(X((t + 1))|X(t), \theta)$ is known but the distribution of the stochastic process in equilibrium, i.e., $f(X|\theta)$ is, except for very special cases, intractable, hence unknown. A further point to note is that the data Y has been assumed to be observed when the underlying process is in equilibrium. In other words, the data is not collected dynamically over time. They refer to such specification as a latent equilibrium process (LEP) model. In the next step, Bhattacharya et al. (2007) developed a methodology to fit LEP within the Bayesian framework. This type of fitting is usually done via the Markov chain Monte Carlo (McMC) method. It is demonstrated by these authors that this kind of fitting is far from straightforward. They developed a different methodology, so as to implement this kind of situation.

Intuitively, we can understand the main idea as follows. Let

$$X_n; \quad n = 1\ldots N;$$

be a Markov chain where the transition $[X_n|X_{n-1}]$ is known but the equilibrium distribution of X_n is unknown. Under certain conditions, X_n converges to a unique stationary distribution with unknown form, and thus the system develops an estimation inference within the Bayesian framework. It can be shown that this convergence-to-equilibrium distribution happens after a time period known as 'convergence time' with the assumption that equilibrium is defined dynamically, as the result of transitions, arising under f (process state at time $t + 1|$ process state at time t, parameters) with t as the time period or generation. There is an important difference between this approach and a dynamic model where input is being continuously addressed, dynamically in time and is collected only when the underlying latent stochastic process is in equilibrium.

In the case of decision making by the brain, the underlying feedback process can be thought of as a stationary stochastic process owing to the presence of noise in the sensory input. It needs a small time period (owing to its discrete nature) to reach equilibrium which is the convergence time mentioned above. However, in the case of infinitely large N, the equilibrium will be reached, irrespective of this convergence time, and the movements may be considered as continuous.

References

Arbib M.A.(2007) Clinical Neuropsychiatry 4(5-6): 208-222

Berstein, N.A. (1967); *The coordination and regulation of the movement*; Oxford, U.K., Pergamon Press.

Bhattacharya, S., Gelfand, A.E. & Holsinger, K.E. (2007); Stat. Comput.;17, 193-208.

Clark A, Chalmers D (1998). Analysis; **58**:7–19.

Clark, A. (2008); *Supersizing the Mind; Embodiment, Action, and Cognitive Extension*, Oxford University Press.

Cox, R.T. (1946) Am.Journ.Phys., 17, 1-13.

Diedrichsen Jörn et al (2009) A probabilistic MR atlas of the human cerebellum; doi:10.1016/j.neuroimage.2009.01.045

Die d Richen, Reza Shadmeher et al; (2002); TICH,834.

Emauael, T & Jordon, I. Michael (2002); Nature Neuroscience 5, 1226 – 1235.

Erimaki, S. & Christakos, C. N. (2008); Coherent motor unit rhythms in the (6-10 Hz) range during time-varying voluntary muscle contractions: Neural mechanism and relation to Rhythmical Motor Control,Neurolophysiol,**99**, 473-483.

Evan Thompson and Mog Stapleton (2008) Making Sense of Sense-Making: Reflections on Enactive and Extended Mind Theories; Springer online, 20 December.

Farmer, S. F. (1998). Journal of Physiology 509(1), 3-14; ibid; Farmer, S,F. (2004); Control of human movement; The journ. of Physiology, **517**(1); Article first published online 8 Sep.; (2004).

Freedman, D.A. (1995); Some issues in the foundation of statistics; Found.Sci., **1**, 19-83.

Goldman, MS (2000); "Computational Implications of Activity-dependent neuronal processes; Harvard Univ"; Ph.D. Thesis.

Goodman, D.A. & Kelso, J.A.(1983); Exploring the functional significance of physiological tremor: a biospectroscopic approach; Exp. Brain Res., **49**, 419-431.

Greene, P.H. (1982); Why it is easy to control your arms?; Journ. Motor Behav., **14**, 260-286.

Grillner Hamburger Victor (1963) Quarterly Rev. of Biol.,**39**, 342-365.

Heijden, J. Van der, Shedmehr Reza; (1992); TICS-834.

Helmholtz, H. Von (1855/1903); *"Ueber das Sehen des Menschen"*, in Vortra « ge und Reden (Braunschweig: Vieweg); 85 – 117; ibid; Helmholtz, Von Hermann (1867); *"(Handbuch der physiologischen Optik"*, Leipzig: Voss: *"Treatise on Physiological Optics"* (1920-25);ibid; (1878/1903); *"Die Tathsachen in der Wahrnehmung"*, in Vortra «ge und Reden (Braunschweig: Vieweg), 213 – 247.

Hommel, B., Jochen Müsseler, J. et.al. (2001); Behavioral and Brain Sciences; **24**, 849–937.

Jaynes, E.T. (1986) *Bayesian Methods; General background*: Cambridge. U.K., Cambridge University Press.

Joliot J.M., Ribary U., & Llinás R. (1994) Proc. Natl Acad. Sci.; USA. 1994; **91**:11;748–751. (doi:10.1073/pnas.91.24.11748).

Kondepudi, D. & Prigogine, I. (1998) *Modern Thermodynamics*, Wiley, New York.

Körding, K.P, Tenenbaum, J.B & Shadmehr, R. (2007); Nature Neuroscience, **10**, 779-786.

Körding, K.P. (2007); Review, Science, **318**, 606-610.

Klagge, J.C. (1999); Wittgenstein on Non-mediative causality; Journ. History of Philosophy; **37** (4), 653-667.

Livesey, P. J. (1986)*Learning and Emotion: A Biological Synthesis; Vol.1of Evolutionary Processe*; Lawrence Erlbaum Associates, Hillsdale, NJ.

Llinás, Rodolfo (1987); *Mindness as a functional state of the brain: in "Mind waves"*; (eds.) C. Blackmore & S. A. Greenfield (Eds.), pp. 339–358). Oxford, England:

Llinás R. R. (1988) Science. 242; pp.1654–1664.

Llinas, R., Leznik, E. & Makarenko, V. (2004) *IEE. J.* **29**; pp. 631-639.

Mackay, D.J.C. (2003); *Information Theory, Inference and learning algorithms*; Cambridge Univ. Press.

Makarenko, V. & Llina's, R. (1998) *Proc .Natl. Acad. Sci. U S A*, **95**, pp. 15747–15752.

Markus F., Peschl, M.F. & Riegler, A.(1999); *Understanding Representation in the Cognitive Sciences*; (Eds: A. Riegler, M. Peschl, & A. von Stein); Kluwer Academic/Plenum Publishers, New York, 1.

Maturana, H.R. & Varela F.J. (1980); *Autopoiesis and cognition: The realization of the living*, Springer.

Maturana, U. (2000); *Steps to an Ecology of Mind*; University of Chicago Press, Chicago.

Moshe Bar (2009) *Predictions: a universal principle in the operation of the human brain*; Trans. of Royal Society B; Royal society publishing org. 30March 2009. DOI: 10.1098/rstb.2008. 0321

Neumann J. Von (1996); *Mathematical foundation of Quantum Mechanics;* Translation from German edition.

Pellionisz, A. & Llinás, R. (1979 ; Neuroscience, **4**, 323-348.

Pellionisz, A. & Llinás,R. (1985) Neuroscience, **16**: 245-273.

Perkel D.H. (1990) *in response to How brains make chaos in order to make sense of the world* Neurocomputing 2 Ed by J.M Anderson et al MIT Press, Cambridge, MA, p. 406

Prigogine, I. & Kondepudi, D. (1998): Modern Thermodynamics: From Heat Engines to Dissipative *Structure*; Wiley, Chichester.

Rosenblatt, Julio, K. (2000) Autonomous Robots; **9**(1), 17-25, Kluwer Academic Publishers.

Roy, S. & Llinás, R. (2008) *Dynamic Geometry, Brain function modelling and Consciousness*; Eds. Banerjee, R. & Chakravorty, B.K.; Progress in Brain research; **168**, 133 (ISSN-0079-6123);

Roy, S. & Llinás, R. (2012) *Role of Noise in Brain function;* Science; Image in action; eds. Z. Bertrand et al., World Scientific Publishers, Singapore; pp 34-44.

Ruiz-Mirazo, K. Moreno, A. (2004) Artificial Life, 10:235–259.

Stanek E., Cheng S. (2014); Neuroscience; **1;**145-59.

Stanek IV, E., Steven Cheng, Jun Takatoh, Bao-Xia Han, and Fan Wang, (2014) Monosynaptic Premotor Circuit Tracing Reveals Neural Substrates 4 for Oro-motor Coordination; Elife (2014).

Tanji, J. & Shima, K. (1994) Nature; **371**:413-16.

Thompson E. (2007) *Mind in life:Biology, Phenomenology, and the sciences ofMind* (pp 4-47). Harvard University Press, Cambridge, MA.

Valbo, A. B. & Wessberg, J. (1993); Journal of Physiology; **469**, 673—691.

Varela, A.B. et al (1974) Biosystems; **5**(4), 187-195. (Elsevier).

Varela, F.J. (1979) *Principles of Biological Autonomy* (New York: Elsevier North Holland).

Varela, F.J. (1997) Brain Cognition, **34**:72–87.

Wessberg, J. & Valbo, A. B. (1995); Journal of Physiology; **485**, 271—282;

Wessberg, J. & Valbo, A. B. (1996); Journal of Physiology; **493**, 895—908.

Wessberg, J. & Kakuda, N. (1999); Journal of Physiology; **517**, 273—285; ibid; Wessberg, J. & Kakuda, N. (2004); Journal of Physiology; **517**; issue **1**; published on line, 8 Sep.

Wheeler, M. E., Paterson, S.E. (2008); Journal of Conitive Neuroscience;**20**(12), 2211-2225.

Wittgenstein, L. (1913) *On logic and How to do it? (original);* The Cambridge review, Vol **34**.

Wittgenstein, L., Klagge, J.C. et al (1997); Cátedra;**41**.

Wittgenstein L., (2001) *Biography and Philosophy*; Cambridge University Press.

Yonas A. (2003) *Development of space perception*, in *Encyclopaedia of cognitive science.*: [ed. Anand R.], Macmillan; London, UK: pp. 96–100.

Chapter 4
New Empirical Evidences on Decision Making and Cognition

Abstract Recently, series of experiments have been performed with human subjects where the law of addition of probabilities in classical probability theory has
been shown to be invalid within the context of decision making in the cognitive
domain. These results are classified into six different categories. The concept of
quantum probability has been introduced to explain the data, but so far, no quantum
mechanical framework has been proposed at the anatomical level of the brain.
Quantum probability is used in the more abstract sense without considering the
concept of elementary particles or the Planck constant, etc. In a sense, this concept
of quantum probability can be used in any branch of knowledge. However, it is
necessary to understand how it can be contextualized in the case of the neuronal
architecture of the brain. It is worth mentioning that the non-commutative structure
in the quantum paradigm has been shown to be valid in the visual architecture of
human brain. The uncertainty relation similar to Heisenberg uncertainty relation has
been found to operate in the visual cortex. This sheds new light on understanding
the data found in the case of ambiguous figures within the above six categories.

Keywords Uncertainty relation · Disjunction effect · Non-commutativity ·
Ambiguous figures · Contextualization · Quantum probability

In modern neuroscience, the concepts that have been developed and have prevailed
until now deal with the problems regarding the extent of the role of the network
responsible, as well as predominant, for performing cognitive functions. But individual neurons are considered as less probable to take a primary part at the network
level, as they are incapable of forming the simple, basic elements of these networks.
This concept is considered by Arshavsky (2003) who states that the configuration of
outstanding cognitive abilities results from individuals of outstanding capabilities.
But the most important part of the work is the introduction of the role of human
behavioral genetics, based on their experimental data. They considered this role as
one of the most probable causes behind the disparities between the cognitive
capabilities of individuals. The central hypothesis behind this is the idea that the
involvement of the neurons, which form the networks in cognitive processes, are

complex. Their functions are not limited, i.e., not just only to the role of generating electrical potentials and releasing neurotransmitters. They suggested that favorable configurations of the micro architecture are responsible for this kind of function, and it exists in cognitive-implied networks. Not only that, but this fact assists in the final formation of ontogenesis which may be occurring in a relatively random fashion. They proposed the concept about the lucky combinations of specific genes that are supposed to determine the intrinsic properties of neurons, involved in cognitive functions of an individual's brain. However, cognition has its own typical characteristics, playing different kind of roles, e.g., reflecting uncertainty together with several possible kinds of involvements; possessing its own typical characteristics in reflecting uncertainty; and containing those characteristics of uncertainty, responsible for both the future events as well as that of the internal states.

All of these factors produce, ultimately, the inevitable consequences based on the facts that life events are often agglomerations of pleasant and unpleasant components (Williams and Aaker 2002). To understand how the cognitive system resolves affective uncertainty presented challenges to White et al. (2013) and many others (Aerts 2009; Atmanspacher et al. 2006). They (Aerts 2009; Atmanspacher et al. 2006) proposed an ambitious new perspective on this question, based on quantum probability (QP) theory (note that in this work, by QP theory, we simply mean only the rules for how to assign probabilities to events from quantum theory). Classical approaches must be assumed to possess particular state, even if knowledge of this state is uncertain. Such an assumption seems straightforward. Otherwise, it would be difficult to build a model. Yet, there is an alternative, intriguing possibility, that emerges from the recent uses of QP theory in cognitive modeling (Busemeyer and Bruza 2012; Pothos and Busemeyer 2013). This theory is based on the framework which is based on the criteria for assigning probabilities to observables and, therefore, potentially relevant wherever there is a need to formalize uncertainty. QP cognitive models often have the same intentions (Griffiths et al. 2010; Oaksford and Chatter 2007) as classical probability models. But, classical and QP frameworks are founded on different axioms. QP models incorporate certain unique features, such as superposition and the capacity for interference, and there has been growing interest in exploring the relevance of such features for cognitive modelling (e.g., Aerts 2009; Atmanspacher et al. 2004; Blutner 2009; Busemeyer et al. 2011; Khrennikov 2010; Pothos and Busemeyer 2009; Wang et al. 2013).

The quantum-probability perspective proposes simple, as well as quite surprising but elegant, empirical predictions. Here, often, probabilistic assessment can be described as strongly context and order dependent. Added to this, unlike the classical approach which assumes a particular state, the QP approach is different since it poses a distinction between particular states and superposition states (i.e., indefinite with respect to some specific judgment). Not only that, but the composite system may be a composite and or an entangled system (i.e., impossible to decompose into a simpler system). There are several fundamental but empirical findings which give substantial evidence regarding many of the complex cognitive processes, in which they are observed obeying quantum rather than the classical probability principles. Moreover, the quantum probability perspective proposes simple, as well as quite surprising, elegant empirical predictions. Therefore, it is possible to consider a situation in which affective evaluation is developed over two steps, such that each step involves a stimulus presentation. Classically, it should not matter whether the person is asked to provide an affective evaluation just after the second step or after the first step as well. So, the intermediate evaluation would simply 'read off' the existing state, and this should not affect the overall outcome of the affective evaluation. However, in the QP model, an action of affective evaluation (a "measurement") can have a profound impact on the state of the system and, therefore, the intermediate evaluation influences the eventual outcome of the second evaluation. Note that a classical model could incorporate the possibility that an evaluation (or rating, etc.) has a constructive role, but this could only be done with additional assumptions, not as part of classical probability theory.

Recently, Aerts et al. (2009), Busemeyers and Trueblood (2011) and others performed a series of detailed experiments, using humans as subjects. They observed that, in plenty of cases, there exists a clear violation of traditional probability theory. An intensive literature survey also indicates the failure of classical probability theory in explaining human cognition modeling, beyond a certain limit.

However, it is worth mentioning that, though the Bayesian approach may appear to be an efficient and promising tool to solve this problem, the complete success story of Bayesian methodology is yet to be established. The substantial presence of epistemic uncertainty seems to pose problems and its eventual effects on cognition, at any instant. The results of the above mentioned experiments can be classified into following six groups as a cognition spectrum of the human mind:

1. Disjunction effect
2. Categorization–decision interaction
3. Perception of ambiguous figures.
4. Conjunction and disjunction fallacies
5. Overextension of category membership
6. Memory recognition over-distribution effect; Fallacies over-distribution effect.

We explain these effects briefly in the following section.

4.1 Disjunction Effect

Amos Tversky and Shafir (1992) discovered a phenomenon called the 'disjunction effect', while testing a rational axiom of decision theory. It is also called the 'sure thing principle' (Savage 1954). This principle states that, if action A and B are chosen which belongs to the state of the world X, and the action A over B is preferred, then, under the complementary state of the world $\sim X$, again, action A over B is preferred, it is expected that one should prefer action A over B even when the state of the world is not known. Symbolically, this can be expressed as:

$$\text{If, } P(A \cap X) > P(B \cap X) \quad \text{and} \quad P(A \cap X^C) > P(B \cap X^C)$$

Then,

$$P(A) = P(A \cap (X \cup X^C)) > P(B \cap (X \cup X^C)) = P(B)$$

That is, $P(A) > P(B)$ is always true. With the aim of testing this principle, in their experiment, Tversky and Shafir (1992) performed the test considering a two-stage gamble by presenting it to 98 students. They adopted a two stage-gamble, i.e., it is possible to play the gamble twice. The gamble is done under the following two conditions:

- The students are informed that they lost the first gamble.
- The students remained unaware of the outcome of the first gamble.

The gamble to be played had equal stakes, i.e., of winning \$200 or losing \$100, for each stage of decision making, i.e., whether to play or not to play the gamble. But the real amount won or lost was, in fact, just \$2.00 or \$1.00 respectively.

Interestingly, the results of these experiments can be described into the following manner:

1. Of the students who won the first gamble, 69 % chose to play the second stage;
2. Of the students who lost, 59 % chose to play again;
3. Of the students who are unaware whether they won or lost, 36 % of them (i.e., less than a majority of the students) chose to play again.

Explaining the findings in terms of choice based on reasoning, Tversky and Shafer (1992) raised some questions about these surprising results: whenever, the students knew that, if they win, they would have extra house money, then they can readily play again, because if they lose, they can play again to recover the loss. For regarding the students who did not know the outcome, then the main issue is *"why does a sizable fraction of the students want to play the game again, since either they won or lost and the result cannot be anything else?"* Thus they arrived at the key result, but faced the problem of explaining the outcome as either a win or a loss. Busemeyer et al. (2006) originally suggested that this finding could be an example of an interference effect, similar to that found in the double-slit experiments conducted in modern particle physics.

Let us consider the following analogy between the disjunction experiment and the classic double-slit type of experiment in physics:

Both the cases involve two possible paths: here, in the disjunction experiment, the two paths are inferring the outcome of either a 'win' or 'a loss' with the first gamble; for the double-split experiment, the two paths are splitting the photon into the upper or lower channel applying a beam splitter. The path taken can be known (observed) or unknown (unobserved), in both the experiments. Finally in both the cases, the fact is that when the case of gambling for disjunction experiment and hence, the detection at a location for the two-slit experiments are considered for the chosen unknown, i.e., unobserved conditions, the resultant probability, meant for observing interference phenomena, is found to be much less than each kind of the probabilities which is observed for the known (observed) cases. Under these circumstances, we can speculate that during the disjunction experiment, under the unknown condition, instead of being definite, so far as the win or loss state is concerned, the student enters a superposition state. In fact, this state prevents finding a reason for choosing to gamble. In double-slit experiments, the law of additivity of probabilities of the occurrence of two mutually exclusive events (particle aspect or wave aspect) is violated, i.e., the total probability is:

$$P_{AB} \neq P_A + P_B$$

for two mutually exclusive events A and B.

This is due to the existence of interference effects, known as the formula of total probability (FTP). It has already been established that the two-slit (interference) experiment is the basic experiment which violates the FTP. Feynman (1951), here and in many of his other works in physics, presented his points with detailed arguments about this experiment. There, the results, i.e., the appearance of interference fringes, appeared to him to be not at all surprising phenomena. He explained it as follows:

In principle, interaction with slits placed on the screen may produce any possible kind of distribution of points on the registration screen. Now, let us try to explain the following quantum probabilistic features that appear only when one considers the following three kind of different experiments (as per Conte et al. 2009):

(a) When only the first slit is open (i.e., the case $B = +1$, in an equivalent manner),
(b) When only the second slit is open (i.e., $B = -1$, in this case),
(c) In the particular case that both slits are open, it is the random variable B determining the slit through which it to passes.

At this stage, let us now choose any point at the registration screen. Then resultant scenario will be as follows:

- the random variable A if $A = +1$
- the opposite case happens if a particle hits the screen at this point, i.e., $A = -1$

Now, for classical particles, the FTP should predict the probability for the (c) experiment (both slits are open), supposed to be provided by the (a) and (b) experiments. But, it has already been mentioned that, in case of quantum particles, the FTP is violated: for the additional cosine-type term appearing on the right-hand side of the FTP, it is the interference effect in probabilities which is responsible. Feynman characterized this particular characteristic feature of the two-slit experiment as the most profound violation of laws of classical probability theory. He explained it the following way:

In an ideal experiment, where there is no presence of any other external uncertain disturbances, the probability of an event, called probability amplitude is the absolute square of the complex quantity. But, when there is a possibility of having the event in many possible ways, the probability amplitude is the sum of the probability amplitudes considered separately. According to their experimental results, Tversky and Shafir (1992) demand that this violation of classical probability is also possible to happen in their experiments with cognitive systems. Though, due to the possible restrictions present in quantum mechanics, we could not begin from Hilbert formalism at the start, this formalism was justified by experiments in the laboratory.

Let us now try to interpret the meaning of the interference effect within the context of the experiments on gambling described above. We will follow, here, Busemeyer's formulation (2011) which is as follows:

Two different judgment tasks A and B are considered in this case. The task A is considered having J (taking $j = 2$, binary choice) different levels of the response variable and B, with K (say, $k = 7$ points of confidence rating) levels of a response measure. For two groups, randomly chosen from the total participants, we have:

- Group A gets task A only;
- Group BA gets task B followed by task A.

The response probabilities can be estimated as follows:

1. From the group A, let $p(A = j)$ be estimated; this denotes the probability of choosing level j out of the response to task A;
2. And then from the group B, let $p(B = k)$ be estimated; the corresponding probability denoted by choosing level k from the task B;
3. Now, it is possible to estimate the conditional probability $p(A = j|B = k)$; which can be stated as the probability of responding with level j from the task A, given the person responded with level k on earlier task B.

So, we can write the estimated interference for level j to task A (produced, when responding to task B) as:

$$\partial A(j) = p(A = j) - p_T(A = j)$$

where, $p_T (A = j)$ denotes the total probability for the response to task A.

In defense of using quantum formalism in the case of human judgements, Busemeyer and Trueblood put forward the four following reasons, beautifully, in their famous paper (Busemeyer et al. 2011):

(a) judgment is not a simple readout from a pre-existing or recorded state, but instead it is constructed from the current context and question; from this first point. it then follows that (b) making a judgment changes the context which disturbs the cognitive system; and the second point implies that (c) changes the context produced by the first judgment which affects the next judgment producing order effects, so that (d) human judgments do not obey the commutative rule of classic probability theory (Busemeyer et al. 2011).

4.2 Categorization–Decision Interaction

Several authors (Townsend et al. 2000) studied the interaction between categorization and decision-making, which gives rise to a new paradigm. In their works, categorization and decision are combined in an experiment where the task is related to photorealistic faces. They assigned two different types of face stimuli, probabilistically, based on the category, to one of two fictitious groups to test Markov models and quantum models. In the next step of the experiment, faces were further probabilistically assigned to categorize whether faces are hostile or friendly (Busemeyer et al. 2009). Participants were shown pictures of faces, on each trial but varied each time, again, along two dimensions (face width and lip thickness). In part I, the participants were asked to categorize probabilistically the faces as to whether each belongs to either a "good guy" or "bad guy" group, and/or they were asked to categorize so as to decide whether to take an "attack" or "withdraw" action, concerning interaction with each. To achieve the results, primary manipulation is produced by following the two-test conditions, presented on different trials, to each participant. In the C-then-D condition, participants made a categorization followed by an action decision and lastly, in the "D-Alone condition", participants only made an action. Undergraduate students from a US Midwestern university participated in the study, and each person participated for six blocks of C-D trials with 34 trials per block, and one block of D-alone trials with the same number of trials per block. C-D trials were pooled across the blocks as very little or no learning was observed due to the supply of all of the necessary information needed. The results of the experiments can be summarized as:

For C−then−D,
$$P(G) = 0.17; P(B) = 0.83; P(A|B) = 0.63; P_T(A) = 0.59; P(A|G) = 0.41$$

And for D−Alone, $P(A) = 0.6$

Then, P_T is the total probability, given by:

$$P_T(A) = P(G) \cdot P(A|G) + P(B) \cdot P(A|B)$$

However, $P(A|G) \leq P(A) \leq P(A|B)$ remains true always. The results of the real experiments, as mentioned above, clearly indicate $P(A)$ substantially exceeds $P_T(A)$. Moreover, after categorizing the face as a "bad guy", the probability of attacking in D-alone condition, i.e., $P(A) > P(A|B)$ is even greater. This suggests that the first part of the results is not an artefact of the presentation of categorization and decision questions within a single trial. Decisions concerning interaction appear to be based on information from the category decision, and not from the face stimuli alone.

4.3   Perception of Ambiguous Figures

The possible presence of interference effects was first investigated by Conte et al. (2009) in connection with the studies on ambiguous figures. They studied this effect intensively following the perceptual domain and found the processes underlying the quantum processes exhibiting interference effects in the quantum nature of the results. They found an extra term depicting so-called quantum interference effects. Later on, experimental studies for the possible existence of the superposition principle were established, specifically, on the foundation of a philosophical, as well as a physical, perspective. These results also establish the presence of quantum mechanical phenomena, even in the cognitive domain.

The need for performing this series of experiments also arose to judge the viability of applying quantum formalism in explaining several phenomena for macroscopic systems, particularly, cognitive systems. In one of the series of experiments, they used ambiguous figures to be analyzed by human subjects in order to verify the viability and the validity of classical probability in their results. The results obtained were always negative due to the finding an extra interference term in the results. Approximately 100 students were part of the experimental groups, divided into two groups chosen at random:

1. Three seconds were allowed for the ones who make a single binary choice (plus vs. minus), concerned with an ambiguous figure A, and
2. Those from the other group were also given 3 s to make a single binary choice when an ambiguous figure B was considered.

As a next step, after a mere 800 ms, the group was asked, after a three-second presentation, for another single binary choice (plus vs. minus) for figure A only. These experiments were conducted on a group of 256 subjects. The results gave proof of evidence that the results indeed show some quantum effect. Thus, their observations suggested that, during perception and cognition of ambiguous figures, mental states appear to follow quantum mechanics. The results of this experiment definitely prove the presence of the interference effect in the cognitive domain. As a next step, when test B preceded test A, for just one type of testing stimulus, the following results were obtained:

1. $p(A+)$ = probability of plus to figure A alone
2. $p(B+)$ = probability of plus to figure B alone
3. $p(A+/|B+)$ = conditional probability of to figure $A+$, given plus to figure B
4. $p(A+/B-)$ = conditional probability of figure $A+$, given minus to figure B
5. $p_T (A+)$: = total probability of plus to figure A and also equivalent probabilities for A_-

Now, remembering the fundamental law of classical probability theory, i.e., the Bayes formula of total probability, we have:

$$P(A = +) = p(B+)p(A = +/B = +) + p(B = -)p(A = +/B = -).$$

The results from the experiments by Conte et al. (2009) are summarized in the following table:

$P(B+)$	$P(A+|B+)$	$P(A+|B-)$	$P_T(A+)$	$P(A+)$
0.62	0.78	0.54	0.69	0.55

Here, the interference effect comes into play as the difference P_T $(A+) - P$ $(A = +) = +0.14$. The law of classical probabilities cannot explain this significant interference effect. Here, the process of image recognition could be considered to be responsible for the characterization of synchronization of firings that are present in the concerned neural network. We may attribute such synchronization as the cause of stabilization to the firings having a fixed frequency. Thus, this case can be considered as a result due to the cause, i.e., the collapse of the wave function. We can consider this quantum-like state of the particular network as characterized by superposition of frequencies, present in neural firings, even before synchronization-collapse. In fact, we could interpret this result as due to the superposition of alternatives, appearing as two ambiguous sub-pictures only, possessing the quantum-like state of the particular neural network which have been denoted by, say, $A(+)$ and $A(-)$, in a given A-figure. Here, we can envisage the quantum-like state of the neural network to be connected to image recognition, as the superposition of two states, i.e., $\varphi(A+) + \varphi(A-)$. In an ideal case, this superposition should be considered as induced by the superposition of neural oscillations on two definite frequencies. But, in the actual situation, each state of $\varphi(A+)$ and $\varphi(A-)$ is found to be realized at the neural level and to have its own range of frequencies. So, we could come to a conclusion that certain quantum-like behavior of the mental states are present in such cases. To explain the above results, the possibility of the presence of an interference term is indicated. These results point to the fact that it is necessary to employ a new kind of approach or alternative model for the proper explanation, in a conclusive manner, of all the results mentioned.

4.4 Conjunction and Disjunction Fallacies

Tversky and Kahneman (1983) studied in detail an important probability judgment in which they found an error called the 'conjunction fallacy'. This can be stated as follow:

A brief story was provided to the judges (e.g., a story of a woman of named Linda. She is a university student and, at the same time, an activist connected to an anti-nuclear movement). Together with this background information, then, the judge is asked to rank the likelihood of the following events:

- She is an active member in the feminist movement (F);
- Event B for the case when Linda is a bank teller;
- Lastly, event $F \cap B$ points to Linda's role, currently, as both connected to the feminist movement as well as a bank teller.

The results are: the conjunction fallacy occurs when option $F \cap B$ is considered to be more likely than option B, even though $F \cap B \subset B$) (Tversky and Kahneman 1983). Carlson and Yates (1989) introduced the term disjunction fallacy. They described a situation where the disjunction $F \cap B$ is less likely than the individual event F. One can obtain the conjunction fallacy using betting procedures where probabilities are not needed, since it appears similar to the negative interference effect caused by possible conjunction error. This observation clearly indicates the violation of classical probability, which can be explained as follows.

Let us define F as the event "yes" to feminism and B denotes the condition when Linda is bank teller, whereas S is for the total Linda story. Now, following the laws of probability:

$$p_{\mathrm{T}}(B|S) = p(F \cap B|S) + p(F \cap B|S) > p(F \cap B|S),$$

but the judgments produce:

$$p(B|S) < p(F \cap B|S) + p(F \cap B|S) < p_{\mathrm{T}}(B|S);$$

which implies a negative interference effect. This observation conforms the necessity of finding a new framework to explain this situation.

4.5 Over Extension of Category Membership

The "guppy effect" can be defined when the membership weights are assigned to a pair of concepts, and their conjunction produces this typical effect. We know that the typical characteristics of a specific item with respect to the conjunction of other concepts can behave in an unexpected way. Sometimes, this type of problem is also referred to as the 'pet-fish problem' and the effect is named the 'guppy effect'. It is observed that subject's rating of cuckoo is better as a member of the conjunction 'Bird and Pet' than for the concept of 'Pet' on its own. If the conjunction of logical propositions governs the effect due to the conjunction of concepts, the second should be at least as great as the first. But this is not so in practice. This deviation is referred to as 'overextension' of this phenomenon, which effect can also be interpreted as the result of the interference effect. Let us define probability $P(A|x)$ in such a way that category A is true, given the item x, and probability $P(A \cap B|x)$ whenever the category $A \cap B$ is true, given the item x. Then law of total probability suggests that $P(A|x)$ is the total probability of A, given the item x. Several other examples can be stated which reveal that this effect exists in abundance.

Thus, as this kind of effect occurs due to the conjunction of more than one concept, a more holistic approach is needed to explain, more clearly, the overextension of category membership.

4.6 Over-Distribution Effect in Memory Recognition

Another test called the conjoint-recognition paradigm test can be traced to the research connected with memory recognition. In order to establish this test, a set of memory targets T is chosen in which the participants are to be rehearsed. This, in fact, means that each participant of this set can be considered as a short description of an event. Hence, a test phase starts after a delay, known as the recognition test phase. During this test phase, the participants are presented with a series of test probes that consists of:

- Trained targets from T
- Non-trained targets from a set R (called a set of distracting events) each member of which is a new event having some meaningful relation to the set of the target event
- Lastly, an unrelated set U consisting of non-target items, each member being completely unrelated to the targets.

Let us start with a memory test trial having a test probe from the set T. The conditional probability $P(V|T)$ is defined as the probability of accepting the target in case the verbatim question is asked. $P(G|T)$ is defined as the probability of accepting the target as gist distractors (G), i.e., requiring acceptance only to distractors which are related to non-targets from R. The third one consists of accepting an instruction from any of previous two categories, i.e., verbatim or gist items (V or G); plus, it is also necessary to accept the related probes from either T or R. Hereafter:

- V represents the event accept as a target from T.
- G is meant for the event accept as a non-target from R.
- And Vor G denotes the event accept as either a target from T or a non-target from R.

Now let us consider the memory test trials employing a test probe, which belongs to the target set T. If the verbatim question is asked, formally, we can define $P(V|T)$ as the conditional probability for accepting the target; when the gist question is asked, the probability of accepting the target is formally denoted by the probability $P(G|T)$; finally when the verbatim or gist question is asked, we denote the probability $P(\text{Vor } G|T)$, formally. Again:

$$P(\text{Vor } G) = P(V|T) + P(G|T)$$

because, obviously, a probe x comes either from the set T or from the set G exclusively, but never from both the sets. Now, it is possible to express a kind of episode over the distribution effect which can be expressed by:

$$\mathrm{EOD}(T) = P(V|T) + P(G|T) - P(V\mathrm{or}\,G|T)$$

In fact, it is often observed that, though recognition probes appear familiar, their presentation is not possible to recollect. Brainerd and Reyna (1988) estimated EOD for 116 different experimental conditions. Dual-process models have been applied in their experimental procedures by which they arrived at the following conclusion:

The results of attributions lead to too many possible presentation contexts, or, in other words, to multiple contexts. These contexts are typically mutually contradictory, named as the phenomenon of **episodic over-distribution**. In the case of process-dissociation paradigms and the conjoint–recognition paradigms–measurement was taken for the attributions to two contradictory contexts. These are as follows:

(a) Presented and not presented, i.e., conjunction recognition
(b) Either presented on list 1 only or presented on list 2 only.

They obtained the following results, consistent with dual-process models but inconsistent with one-process models. After the completion of the analysis of over 100 sets of conjoint–recognition data, the results revealed the presence of the 'attribution of probes' in the first contradictory combinations, which appears virtually universal. Across the data regime, 18 % of studied targets have been noted with true memory probes and 13 % of related distractors observed in false-memory probes. These results were judged as "have been both presented and not presented". It has been found that, across the data collected, 10 % of 116 conditions produced this effect with the mean value of EOD equals 18.

All these six categories of findings raise important issues about whether a classical probabilistic paradigm like the Bayesian framework is a valid one to explain these findings consistently. The alternate framework called quantum probability seems to be a promising one.

4.7   Failures of Commutativity in Decision Making

The concept of commutativity plays a crucial role in understanding quantum logic and quantum probability. Many authors (DeValois et al. 1988; Tourangeau et al. 1991) have pointed out and observed the failure of commutativity in decision making, when asking the same two questions but in different order. They placed two questions: "Is Clinton honest?" and "Is Gore honest?" in reverse order compared to the first arrangement of those same two questions. The results of those two sets of questioning came as a surprise. When these first two questions were asked in a Gallup poll in one order, the probabilities of answering "yes" for Clinton and

Gore were 50 and 68 % respectively. But, surprisingly, the probabilities found were 57 and 60 % (Moore 2002) respectively when the same questions were put to them in the reverse order. Failures of such a kind of ordering, or mathematically speaking, the failures of commutativity cannot be explained by classical probability (CP) theory, because it is to be noted that the probability of saying "yes" to question A and then "yes" to question B

$$= \Pr ob(A). \Pr ob(B \cap A) = \Pr ob(A \cap B) = \Pr ob(B \cap A) = \Pr ob(B). \Pr ob(A|B)$$

Therefore, the prediction of CP theory is that the order of asking two questions does not matter. But in contrast to this, in social psychology, it is possible to explain this order effect in the following manner: the first question activates thoughts which subsequently affects the considerations of the second question (Schwarz et al. 2007). In this way, QP theory can accommodate order effects in Gallup polls in a way which is analogous to the way the conjunction fallacy is explained. The problem of non-commutativity has been studied extensively by Conte et al. (2009), taking a realistic example, i.e., studying the ambiguity of figures in the visual domain where this property turns out to be a prominent one. In the next section, we will discuss this problem taking realistic examples of non-commutativity in the context of the visual cortex.

4.7.1   Non-commutativity and the Uncertainty Principle Underlying the Functional Architecture of the V1 Cortical Area

It is a well-known fact that Heisenberg (1969) introduced the uncertainty relation in quantum physics. He applied this principle in various fields of physical phenomena to quantify the role of non-commutative measurements. Since the famous works of Gabor (1946) and his interpretation of the uncertainty relation, especially in signal processing, this phenomenon has drawn serious attention from the community dealing with signal processing. This principle has been valued by the community as a central tool whenever the need arises, in the topic of coherent states, as well as in optimal measurements. Moreover, this principle can be stated in very general terms on Hilbert space for pairs of non-commuting, self-adjoint operators. In particular, it is to be noted that this principle has also been applied in image processing.

However, immediately after the publication of papers by Wilson and Knutsson (1988) and Daugman (1985), many researchers, who had been dealing with this kind of problem in computer vision, started to employ this approach in a serious manner. In his pioneering work, Daugman explained the features of the VI cortical area. To do this, he analyzed the psychophysiological evidence of receptive fields in the mammalian visual cortex. His works, published in 1985, clearly indicated the presence of constraints in two-dimensional spatial linear filters, where he pointed to the presence of general uncertainty relations that are responsible for attaining spatial frequency, two-dimensional (2D) spatial positions and also the necessary information resolution needed for orientation. From his critical observation, it has been shown very clearly that, though these two quantities in simple cells try to localize position and frequency simultaneously, they do not commute. To calculate the theoretical lower limit for the joint entropy, the so-called uncertainty of these variables, he employed an optical 2D-filter family. His studies reached the conclusion that it is not possible to detect both position and momentum with arbitrary precision, as pointed out in the uncertainty principle.

In recent analysis, Barbieri et al. (2010), while dealing with the generators of the Lie algebra connected to rotations and translations, specifically tried to minimize the role of the uncertainty principle. He explained in his work in detail how the SE (2) group structure is able to provide a different source of non-commutativity in order to explain the features of the VI cortical area. The non-commutativity of the group structure can be explained and proved from fact in the following manner: i.e., in the case of the two transformations, applied in the reverse order, a translation Y_1, followed by a rotation Y_2, does not act in a similar way. It is the treatment which makes the difference in the increment, i.e., the orthogonality to the direction along which the translation occurs makes the difference. Whenever gradients of activity are studied along the integral curves, the results have direct consequences on the property involved. In fact, Lie derivatives are infinitesimal transformations lying along the integral curves of the structure. Here, the non-commutativity of the group can be implied to the non-commutativity of derivatives. Computing the second mixed derivatives we have:

$$Y_2 Y_1 f = (\tilde{y}\partial_{\tilde{x}} - \tilde{x}\partial_{\tilde{y}})\partial_{\tilde{x}} f = \tilde{y}\partial_{\tilde{x}\tilde{x}} f - \tilde{x}\partial_{\tilde{y}\tilde{x}} f$$

Simple algebraic calculations give:

$$Y_2 Y_1 f \neq Y_1 Y_2 f$$

Here, the non-commutativity of Y_1 and Y_2 can be defined as

$$[Y_1, Y_2]f = Y_2 Y_1 f - Y_1 Y_2 f.$$

This is known as the measure of non-commutativity of the space, and the bracket is known as a Lie bracket. Now, if we denote

$$Y_3 = [Y_1, Y_2] = \partial_{\tilde{x}}$$

Similar properties can be defined in the same way for the operators $\hat{Y}_1$ and $\hat{Y}_2$ in a Fourier space. Then, the commutation relation becomes:

$$\hat{Y}_3 = \hat{Y}_2\hat{Y}_1\hat{f} - \hat{Y}_1\hat{Y}_2 f - i\rho_0 \cos(\phi)\hat{f}$$

These two non-commutating operators are considered as position and momentum in quantum mechanics. Under specific condition, i.e., in the case of normalized functions, the variances of the position and momentum operators have a lower bound. This fact has been formalized by Heisenberg in his famous uncertainty principle. This is given by

$$\|xf\| \|\partial_x f\| \geq \frac{1}{2}$$

where, $\|.\|$ stands for the usual L^2 norm.

Before discussing the uncertainty relation and non-commutating operators in more detail, let us describe the functional architecture underlying the area VI of the visual cortex in the brain.

4.7.2 Architecture of VI Area

Hubel and Wiesel showed in their famous works (Hubel and Wiesel 1977) that, in the V1 area of some species, including humans, an array of modules called orientation-preference columns is responsible for the processing of visual information. It is called pinwheels because, like the spokes of a wheel, the orientation columns have been found to be radially arranged around singular points. Bonhoeffer and Grinvald (1991) were the first who observed this pinwheel structure by applying optical imaging techniques. Recently, Ohki et al. (2006) learned the organization of this area with single-cell precision by applying in vivo two-photon imaging. In fact, this kind of hypotheses had been proposed by many who tried to establish the morphogenesis of pinwheels, stating that pinwheels are responsible during visual development and can claimed as de facto optimal distributions for the purpose of coding and availing the most accurate angular position and momentum. Many of studies (Ben-shahar and Zucker 2004; Zweck and Williams 2004; Citti and Sarti 2006; Bressloff et al. 2001; Franken and Duits 2009; Duits et al. 2007; Sarti et al. 2008) done over the last decade, recognized that there exists local invariance in the functional architecture of V1, with respect to the symmetry group of rotations and translations SE(2). Bonhoeffer and Grinvald (1991) measured the orientation of cortical maps in order to construct pinwheels. Again, Barbieri et al. (2010) discussed other possibilities for tackling this problem. They showed that it can be modelled as geometric configurations which are coherent with the states, so that it can be localized as well as best fitted by considering both the angular position and momentum. The theory, adopted by them is based on the well-known uncertainty principle. It is well known that after the discovery of the Heisenberg uncertainty principle in quantum mechanics (Cohen-Tannoudji et al. 1978), many researchers used it by applying it to invariance-related problems (Folland 1989;

Carruthers and Nieto 1968; Isham and Klauder 1991). With the applications of this classical principle, it has been possible to define the states having uncertainty between the two non-commutative quantities, i.e., of linear position and linear momentum.

Daughman (1991) adopted this principle in modeling the receptive profiles of simple cells. He considered these profiles as minimizers, defined on the retinal plane. The minimizers of the uncertainty are defined as coherent states which correspond to the functions with optimal localization, both in angular position and angular momentum. The fact that the cortical geometry follows the same corresponding principle and can be described with SE(2) is based on the uncertainty between the two non-commutative quantities, i.e., of angular position and angular momentum. Though these principles have been proposed by many (Wolf and Geisel 1998; Bressloff and Cowan 2003) and many have reviewed this problem (Swindle 1996; Mortimer et al. 2009), it remains open to study further.

In their basic works, Hubel and Wiesel (1962) clearly pointed out that the primary visual cortex consists of hypercolumns of simple cells. These cells are sensitive to the position (x_0, y_0) and the local orientation θ of the stimulus. Neurons in the striate cortex are organized into columns and hypercolumns, i.e., "*With each column containing neurons with similar properties and each hyper column processing one small section of the visual world. This activity is designed to help you understand how columns and hypercolumns work to process the visual world-* "*from internet source* (Wolfe et al. 1969, Activity 3.5)

In this case, in order to explain the functional organization of the visual cortex, a three-dimensional model of hypercolumns has been considered. The group-theoretical analysis points to the fact that this organization can be explained in terms of Euclidean Lie group SE(2), i.e., the continuous group of translations (x_0, y_0) and rotation θ on the plane. Here:

$$SE(2) \approx R^2 \times S^1$$

is taken as a description of the functional architecture. The Lie algebra of SE(2) is used in this case to describe the connectivity between hypercolumns, and the corresponding generators are chosen as:

$$\dot{X}_1 = (-\sin\theta, \cos\theta, 0),$$
$$\dot{X}_2 = (0, 0, 1)$$

which are the vector fields. It is to be noted that, though SE(2) is a 3D group associated to the primary visual cortex, its physical implementation is necessary to be realized on a 2D level, provided by the layer. The next issue, then, is to find out the states minimizing the uncertainty relation. It is possible to find a family of states constructed based on geometry of SE(2), which are optimal functions of spatial and angular position. Mathematically speaking, it is possible to model the orientation activity maps as orientation coherent states, which can minimize the uncertainty, due to the presence of functional geometry, underlying the VI area.

4.8 Uncertainty Relation and Ambiguity in Perception

Conte et al. (2009) performed many experiments, especially in connection with cognition of ambiguous figures, in order to find any possibility of the presence of quantum-like behavior in mental states which could be connected to human cognition. The investigations, based on the nature of perception, clearly show that, when an object is mentally represented through visual input, at any instant, it appears unique even if there is awareness of the possible presence of ambiguity, in any given representation.

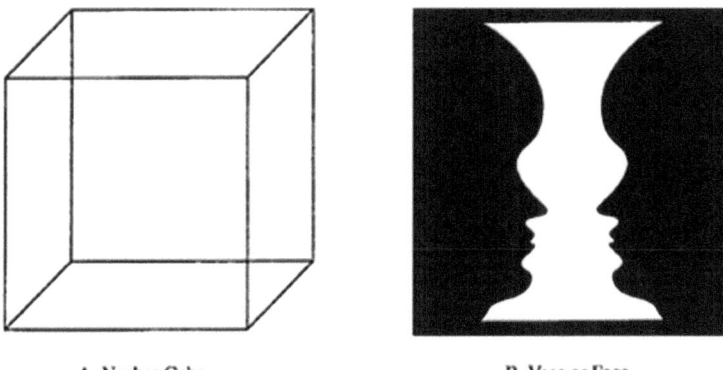

A. Necker Cube   B. Vase or Face

Abrupt visual state changes. (a) Necker cube. (b) Vase/face figure.
Modified from Boring et al. (1948).

In fact, the crucial point to be explained lies in the phenomena that, given the many possible choice of representation, determines which particular representation commands one's attention. **Necker cube** (1832) is one of the well-known examples in this direction—when the cube is seen in one of two ways and at any given instant, only one representation is apparent. It may appear that, perhaps, recent or repeated prior use of one representation plays a role in advantageous positioning of that representation over the other. In such a case, stochastics could be the origin for this behavior, but there remains the possibility of the existence of some underlying factors that might give the edge to one of the representations over the other. This is why the study of ambiguous figures continues to intrigue psychologists and neurologists. Many such research studies have already been done from time to time; various theories have been proposed (Kohler 1940; Gibson 1950; Atteneve 1971; Turvey 1977; Toppino and Long 1987; Horitz and O'Leavy 1993; Accardi 2006).

Borsellino et al. (1982) made observations to find the effects of visual angle on the reversal of ambiguous figures. He studied the reversal rates of an ambiguous figure (Necker cube) for different pattern sizes covering a range of visual angles $\emptyset$, from $\sim 1°$ to $62°$. In such cases, it is crucially important to find out the optimal state associated to the VI cortical area that minimizes the uncertainty relation involving angle. This will lead us to build up a comprehensive model for understanding the ambiguous figures and its link to quantum-like behavior in cognitive domain.

4.9 Uncertainty Relations for Unsharp Observables

Recently, Cohen (1966, 1989) systematically studied the time-frequency analysis of signals and the correspondence between signal analysis and quantum mechanics. He claimed the uncertainty relation to be merely a relationship concerning the marginals only, which has no bearing on the existence of joint distributions; in particular, he raised critical questions about the validity of mathematical correspondence between quantum mechanics and signal analysis. Roy et al. (1996) developed a theory of unsharp measurements in quantum mechanics, which opens up the new possibility of constructing joint distributions of unsharp observables and studying their relevance to the Heisenberg uncertainty principle. This framework is quite interesting not only in quantum mechanics, but also in the context of signal processing, especially, in image processing and pattern recognition. If one considers the case of object-background segmentation in a grey-level image, a great deal of ambiguity arises in the decision process due to the continuous grey level-distribution from object to the background. This kind of situation could be handled by fuzzy set theory instead of usual set theory.

Busch and Lahti (1984) questioned the exclusive validity of a statistical interpretation of uncertainty relations. They tried to justify an individualistic interpretation by introducing the concept of unsharp observable where a joint measurement of position and momentum is a logically tenable hypothesis. It is the transition from ordinary sets to fuzzy sets that enables one to describe this type of unsharpness. In ordinary set theory, the element relations x $\in$ E can be represented by a characteristic function X_E which is either 0 or 1, while in case of a fuzzy set, E_x is replaced by a more general function μ_E with the following properties:

$$0 \leq \mu_E \leq 1$$

Here, the main difference from a characteristic function X_E is that μ_E may assume any value in the real interval [0, 1]. Now we proceed with the corresponding change in the description of observables. A spectral projection $Q(E)$ is defined in the configuration space representation by means of the equation:

$$(Q(E)\phi)(q) = X_E(a)\phi(q)$$

Replacing X_E by μ_E, we get the new effect:

$$(a(E)\phi)(q) = \mu_E(q)\phi(a)$$

where $a: E \rightarrow a(E)$ is a positive-operator-valued (POV) measure called an unsharp position observable or a fuzzy observable.

The analysis made on the basis of unsharp observables by Roy et al. (1996) has shown that, if a system possesses position and momentum only with indetermination to the extent given by the probabilistic uncertainty relations, then joint measurements should be impossible with accuracies violating the uncertainty relation. With the introduction of the so-called unsharp observable, a proper joint probability distribution for position and momentum has been developed. It is a well- known fact that, according to the conventional interpretation of uncertainty relation, it is not possible to determine the two complementary aspects simultaneously. However, according to unsharp observables, it is possible to determine these complementary aspects simultaneously, but not with infinite precesion. The price one has to pay is to have unsharpness in measuring the observables, which satisfy a inequality like:

$$\Delta f \cdot \Delta g = \Delta|\psi|^2 \frac{h}{2\pi} \Delta|\bar{\psi}|^2 \geq \frac{h}{4\pi}$$

so that the inaccuracies or unsharpness Δf and Δg of the Fourier couple (Q_f, P_g) obey the uncertainty relation. In visual information processing, there are several issues which have important bearing on quantum-mechanical formalism. This implies that any operation, whatever, upon spatial data, gives rise to ambiguity, reflected in the uncertainty of measured properties in two complementary domains.

In later chapters, we will discuss the quantum probability and quantum ontology in detail. Before going into the details of these discussions, the basic concepts of quantum theory will be discussed in the following chapters so as to explain the concept of quantum probability and quantum ontology in a comprehensive manner. In the first half of twentieth century, after the discoveries based on the double-slit experiment, physicists introduced the concept of quantum probability. The motivation for performing this experiment was to understand the wave–particle duality of a microscopic entity. With this aim, we shall try to give a basic description of the double-slit experiment and wave-particle dualism in the next section.

4.10   Wave-Particle Dualism and Double Slit Experiment

In 1690, Huyghens proposed a theory of light which encompasses all the properties (known at that time) of propagation of light in a medium, assuming the wave property of light. Huyghens (1690) clearly understood that the phenomena of diffraction and interference have natural explanations using the wave theory of light. In 1704, Newton published his theory of light proposing that light travels in straight line, and it consist of time particles called corpuscles–hence, known as the corpuscular theory of light. The debates regarding the corpuscular and wave natures of light continued for more than 200 years before its resolution in quantum theory. In 1905, Einstein explained the phenomena of the photoelectric effect based on the corpuscular theory of light. Briefly, this effect can be stated as that, when lights fall on the surface of a metal, a small current flows. Hence, this effect is named the famous photo-electric effect. Einstein considered light as consisting of particles that force the electrons in the metal to move around. However, it is necessary to consider the wave nature of light to explain the phenomena of diffraction and interference. These experiments, performed in the laboratory, clearly indicate the existence of both wave and corpuscular natures of light depending on the situations. In early twentieth century, these results opened up a new vista in formulating the quantum theory. During this period, the works of Rutherford and Bohr paved the way to construct models of the atom and prove the existence of a small entity like the electron (a negatively-charged entity), proton (positively charged), neutron (neutral), etc. French physicist de Broglie (1924) proposed that the microscopic entity, like the electron, is associated with a wave and the material entity like electron would behave like a wave or a particle depending, on the experimental arrangements.

According to de Broglie hypothesis, the wavelength associated with an entity in the microscopic domain is:

$$\Lambda_{\text{de Broglie}} = h/mv = h/p$$

h being Planck constant and p the momentum of the entity.

Now we have the wave-particle duality for a material entity like the wave-particle duality of the photon. Here, since according to de Broglie hypothesis, the material entity like an electron behaves like a wave, it is possible to test this hypothesis, noting the interference phenomena in an experiment called the "double-slit experiment". Thomas Young demonstrated the wave-like properties of light in an experiment called the "two slit" or "double slit" experiment. This is schematically shown below.

A light wave passes through a screen with two slits of equal diameter and impinges on a screen behind these slits to display the interference fringes. The light behaves like waves. So when they pass through both the slits, a superposition of the waves occurs and causes constructive or destructive interference. This experiment clearly demonstrates the wave nature of light. Similar experiments can be performed to test the de-Broglie hypothesis, i.e., the dual wave and particle aspects of the electron. In this kind of experiment, the electron beam was used instead of a light beam. This experiment is schematically shown as follows.

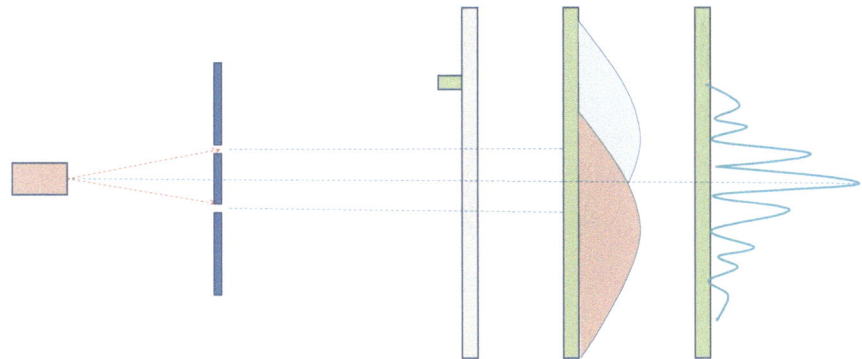

Electrons are fired from the gun along the slits 1 and 2. Behind the slits, detectors are placed to detect through which path the electron goes and the interference fringes appear on the screen behind the detectors. When the slit 2 is closed and slit 1 is open, we get the probability distribution P_1 of the electrons, and, if slit 2 is open and 1 is closed, we get probability distribution P_2. Now, when both the slits are kept open, then the complex curve P_{12} is not the sum of the curves P_1 and P_2. Technically speaking, when the two slits are open, the probability distributions of the electrons are:

$$P_{12.} \neq P_1 + P_2$$

In quantum theory, the probability amplitude of arrival at X, is

$$P_{12} = |\varphi(X)|^2$$

taking,

$$P_1 = |\varphi_1(X)|^2; \quad P_2 = |\varphi_2(X)|^2$$

But, $\varphi(X) = a\varphi_1(X) + b\varphi_2(X)$, because of the superposition principle and, a and b are non-zero constants and $|a|^2 + |b|^2 = 1$

Now, let us consider P_1 and P_2 as:

$$P_1 = |a\phi_1(X)|^2 \quad \text{and} \quad P_2 = |b\phi_2(X)|^2$$

Now, if we watch through which path the electron is passing, only then will $P_{12} = P_1 + P_2,$ but we are not able to watch it, so:

$$P_{12.} \neq P_1 + P_2$$

But why do the chances changes due to our watching? The simple answer is: we use light in this process of watching, and, as light in collision may alter the motion of the electron, hence its chance of arrival may be altered.

In classical probability theory, the total probability can be written as sum of the probabilities of two mutually exclusive events like the particle and wave aspects. However, the double-slit experiments clearly establish the failure of the additivity of probabilities in the case of quantum probability. This is due to the existence of interference term in total probability, i.e.:

$$P_{12} = P_1 + P_2 + {}'\text{Interference}'.$$

This is also due to the fact that classical logic is Boolean in character, i.e., "either or" result, but, in the case of quantum logic, it is non-Boolean in character, because of the intermediate situation due to the superposition effect. So, physical experiments like double-slit experiments launched the shift of paradigms from the classical probability concept to quantum probability and establish the issue of the wave-particle duality of light and matter on solid ground. In other ways, it can be concluded that, for the microscopic entity like the electron, the laws of combining probabilities are different from the laws of probability theory of Laplace. However, it has been shown that the quantum mechanical laws of the physical world approach very closely to the laws of Newton, if only the size of the objects becomes larger. It was Laplace who combined the laws of probabilities in the microscopic world. It is worth mentioning that, during late 60 s of twentieth century, Feynman published a comprehensive article which explains beautifully the concept of probability and the double-slit experiment.

Immediately after the publication of the results about the double-slit experiment for material particles, the debate centered on the concept of wave-particle dualism in the microscopic domain. Different experimental arrangements are thought to be needed to test the wave or particle aspect of an entity in the microscopic domain. Generally, it is believed that wave and particle aspects cannot be detected in a single experiment. Thus, it provokes the epistemological debate about the complementary principle–the microscopic entity sometimes behaves like a wave, sometimes behaves like a particle. In the double-slit experiment, an electron as a particle will pass through either of the holes at one time. So, if one places a detector on either of the paths through which the particle passes, it is possible to detect the path through which the particle passes, but the interference pattern will immediately be washed away due the observation with the detector. In such a case, one can detect either the particle aspect or the wave aspect (manifested as an interference pattern) at the same time. This mutual exclusiveness lies in the heart of Bohr's principle of complementarity (1920). However, the information-theoretic analysis of the double-slit experiment by Wootters and Zurek (1982) demonstrate the simultaneous existence of complementary aspects like **wave** and **particle** in a single experiment. Heisenberg, himself, discussed the possibility of the simultaneous existence of complementary aspects. In our previous chapter, we have discussed the concept of unsharp observables in quantum mechanics. Applying this concept of unsharp measurement, it is possible to prove that the two complementary aspects can be determined simultaneously in the same experiment, but not with infinite precision. We have mentioned also that the unsharpness in measuring the observables satisfies an inequality like:

$$\Delta f \cdot \Delta g = \Delta |\psi|^2 \frac{h}{2\pi} \Delta |\bar{\psi}|^2 \geq \frac{h}{4\pi}$$

So that the inaccuracies or unsharpness Δf and Δg of the Fourier couple (Q_f, P_g) obey the uncertainty relation.

Recent activity by experimenters in Germany led to claims that one could observe both wave and particle properties of light in one experiment(Mensal et al. 2012). This claim seems to violate Bohr's principle of complementarity. Regarding the fair-sampling perspective related to an apparent violation of duality (Bolduc et al. 2014), Bolduc published a paper in August 2014, reanalyzing the experimental results of Menzel et al. (2012). The results show that the **duality principle is not violated** in the above mentioned experiment, i.e., either the wave aspect or particle aspect will survive at any one time.

References

Accardi, L. (2006); "*Could one now convince Einstein? In quantum theory reconsideration of Foundations*"; 3: G. Adenier, et.al. (Eds.); American Inst. of Physics;Ser.Conf.Procd.;**810**, Melville.N.Y.,3-18.

Aerts, D. & Aerts, S. (1994); Foundations of Science; **1**;85-97.

Aerts, D. et al (2009); - PLoS Genet. **5**(1): e1000351. FlyBase Identifier, FBrf0206745.

Aerts, D. (2009) Journal of Mathematical Psychology; **53**(5); 314-348.

Arshavsky Yuri I. (2003); Brain & Mind; **4**(3);327-339.

Atmanspacher, H., Filk, T. & Romer,H. (2004) Biological Cybernetics; **90**(1); 33-40.

Atmanspacher, H., Romer, H., & Wallach, H. (2006); Foundations of Physics; **32**; 379-406.

Atterneve, F. (1971) Scientific American;**225**;62.

Barbieri, D., Citti, G. et al (2010); An uncertainty principle underlying the pinwheel structure in the [math-phys] primary visual cortex: ([arxiv:1007.1395v1,[Math-ph],8 July.

Ben-Shahar, O. & Zucker, S. (2004) Neural Computation; **16**(3);445-476

Blutner, R. (2009); "*Concepts and bounded rationality: An application of Niestegge's approach to conditional quantum probabilities*". In L. Accardi, G. Adenier, C. Fuchs, G. Jaeger, A.Y. Khrennikov, J. Larsson, & S. Stenholm (Eds.), *Foundations of probability and physics- 5* (Vol. **1101**).

Bohnoeffer, T & Grinvald, A. (1991); Nature; 353(6343); 429-31.

Bolduc, E., Leach, J. et al (2014); PNAS, **111**(34), 12337–12341, *Fair sampling perspective on an apparent violation of duality*; Ed. Peter J. Rossky; doi:10.1073/pnas.1400106111

Borsellino, A., Carlini, F., et al. (1982) Perception; **11**; 263–273.

Busemeyer, J.R., Wang, Z., et al. (2006) Journ. Mathematical Psychology; **50**; 220-241.

Busemeyer, J.R., Wang, Z. & Lambert-Mogiliansky (2009) Journ. Math. Psychology; **53**; 423-433.

Busemeyer, J.R. et al. (2011); Psychological Review; **118;** (2); 193-218.
Busemeyer, J. R. & Bruza, P. (2012); *Quantum models of cognition and decision making;* Cambridge, UK: Cambridge University Press.
Busemeyer, J.R., Pothos, E.M. et al (2011) Phys Rev.; **118**(2);193-218.
Busch, P. & Lahti, P.J. (1984) Phys. Rev; **D 29**; 1634.
Brainerd, C.J. & Reyna, V.F. (1988) Journal of Memory & language; **58**; 765-786.
Bresslo, P.C., Cowan, J.D. et al. (2001) Philosophical Transactions of the Royal Society of London; Series B: Biological Sciences; **356**(1407);299-330.
Bresslo, P.C. & Cowan, J.D. (2003); *Philosophical transactions of the Royal Society of London.* Series B: Biological sciences; **358**(1438); 1643-1667.
Carlson, B.W. & Yates, J.F. (1989) Organizational behavior & human decision processes;**44**;368-379.
Carruthers, P. & Nieto, N.M. (1968) Reviews of Modern Physics; **40**(2);411-440.
Citti, G. & Sarti, A. (2006) Journal of Mathematical Imaging and Vision; **24**(3);307-326.
Cohen, Leon (1966); J. Math. Phys.; **7**;781.
Cohen Leon (1989); Procd.IEEE; **77**(7); 941-961.
Cohen-Tannoudji, C., Diu, B. & Laloe, F. (1978); *"Quantum Mechanics"*; John Wiley & Sons.
Conte, E., Khrennikov, A., Todarello, O., Federici, A. et al (2009) Open Systems and Information Dynamics; **16**; 1-17.
Daugman, J.G. (1985) J. Opt. Soc. Am. A. **2**(7);1160-1169.
DeBrioglie, Luis (1924) *Recherches sur la theorie des Quanta* (University of Paris); (Eng translation:. *"Phase Waves of Louis deBroglie"*) (1972); Am. J. Phys.; **40**(9); 1315-1320.
DeValois, R. Feldman, J.M. & Lynch, J.G. (1988) Journ.Applied psychology; **73**(3);421-435.
Duits, R. et al. (2007) Int. Journ. Comp. Vision; **72**(1); 79-102.
Einstein Albert (1905) *Annalen der Physik;* **17;** 132-148.
Einstein Albert (1905) Explanation of the photoelectric effect with use of the quantum hypothesis of Planck. Light is a flow of corpuscular objects with definite energies Planck's quanta of energy. http://einsteinpapers.press.princeton.edu/vol2-trans/100.
Feldman, M. & Lynch, J.G. (1981) Journal of Applied Psychology; **66**; 127-148.
Feynman, R. (1951) *The concept of Probability in quantum mechanics*; www.johnboccio.com/research/quantum/notes/Feynman-Prob.pdf
Folland, G. (1989); *Harmonic analysis on phase space*; Princeton University Press.
Franken, E. & Duits, R. (2009); International Journal of Computer Vision; **85**(3);253-278.
Gabor, D. (1946) J. IEE, London; **93**; 429–457.
Gibson, J.J. (1950); *"The perception of the visual world"*; Boston M A;Houghton Miffin.
Griffiths, Thomas L. et al., (2010) Trends in Cognitive Sciences; **14**(8); 357-364.
Hampton, J. A. (1988)(a) Learning, Memory & Cognition; **16**; 579-591.
Hampton, J. A. (1988)(b) Journ. Exp. Psychology: Learning, Memory, and Cognition; **14;** 12-32.
Hampton, J.A. (2013); *"Conceptual combination: extension and intension; Commentary on Aerts, Gabora & Sozzo"*; Topics in Cognitive Science: Special Issue"; Modelling with Quantum Probability.
Heisenberg, W. (1969); " Der Teil und das das Ganze; Gesprache im Umkreis der Atomphysik". Munich, Germany: R. Piper, 1969.
Horitz, K.L. & O'Leavy (1993) Perception and Psychophysics; **53**; 668.
Huyghens, Christiaan (1690); *Trail de la lumiere.*
Heisenberg, W. (1969); *"The physical principles of the quantum theory';* Dover, New York.
Hubel, D. & Wiesel, T. N. (1962) J. Physiol; **160**; 106–154.
Hubel, D.H. & Wiesel, T.N. (1977) Royal Society of London Proceedings Series B, **198**:1–59.
Isham, C.J. & Klauder, J.R. (1991) Journ. Math. Phys.; **32**(3):607-620.
Khrennikov, A. Y. (2010). *Ubiquitous quantum structure:From psychology to finance*; Berlin: Springer-Verlag.
Kohler, W. (1940) *Dynamics in Psychology*; New York, Liveright.
Menzel, R., Puhlman, D. et al. (2012); Procd. National Acad. Science; **109**(14); 9314-9319.
Mortimer, D., Feldner, J., et al (2009) Proc.Natl.Acad.Sci.,USA,106, 10296–10301

Moore, D.W. (2002) Public opinion Quarterly; 66;80-91.

Necker, L.A. (1832) Philosophical Magazine;3;329.

Oaksford, M. & Chatter, N. (2007); *Bayesian rationality: The probabilistic approach to human reasoning*; Oxford: Oxford University Press.

Ohki, K., Chung, P. Kara (2006) Nature; **442**(7105); 925-928.

Pothos, E. M. & Busemeyer, J. R. (2009) Procd. of the Royal Society B: Biological Sciences; 276 (1665);2171-2178.

Pothos, E. M. & Busemeyer, J. R. (2013) Behavioral and Brain Sciences; **36**(3); 255-274.

Roy, Sisir, Kundu, M. & Granlund G. (1996); Information sciences;**89**;193-209.

Sarti, A., Citti, G. & Petitot, J. (2008) Biological Cybernetics; **98**(1);33-48.

Schwarz, Nobert et al. (2007) Advances in Experimental social Psychology, Vol **39**; Elsevier Inc.

Savage, L. J. (1954); *The foundations of statistics*; Wiley.

Shafir, E., Simonson, I. & Tversky, A. (1993) Cognition; **49**(1-2);11-36; [aHM]

Swindale, N.V. (1996); The development of topography in the visual cortex: a review of models. Network (Bristol, England); **7**(2):161-247.

Toppino, T.C., Long, G.M. (1987) Perception and Psychophysics;**42**;370.

Tourangeau, R., Rips, L. J., & Rasinski, K. A. (2000) *The psychology of survey response*; Cambridge University Press.

Tourangeau, R., Rasinski, K.A. & Bradburn, N. (1991);Public Opinion Quarterly; **55**;255-266.

Townsend, J. T.; Silva, K. M.; et al (2000) Pragmatics and Cognition; **8**; 83- 105.

Turvey, M. (1977) Psychological Review; **84**;67.

Tversky, A. & Kahneman, D. (1983) Psychological Review; **90**(4);293– 315.

Tversky, A. & Shafir, E. (1992) Psychological sciences;**3**;(6);358-361.

Tversky, A. & Shafir, E. (1992) Psychological Science; **3**(5); 305 – 309.

Wang, Z., Busemeyer, J. R., Atmanspacher, H., & Pothos, E. (2013) Topics in Cognitive Science;**5** (4), 672-688.

Wilson, R. and H. Knutsson, H. (1988) IEEE Trans., SMC; **18**; 305.

White, Lee, C. et al (2013); A quantum probability perspective on the nature of psychological uncertainty; Procd. 35th Ann. conf. cogn. Sci. society; 1599-1604.

Williams, P. & Aaker, J.L. (2002) Journal of Consumer Research; **28**(4); 636-649.

Wolf, F & Geisel, T. (1998) Nature; **395**(6697):73-78.

Wolfe M. et al (1969) *Sensation and Perception*, Fourth Edition, Sinauer Associates, Chapter 3, Activity 3.5.

Wootters, W.K., Zurek, W.H. (1982) Phys Rev D; Part Fields; **121**;580–590.

Zweck, J & Williams, L.R. (2004) Journ. Math. Imaging and Vision; **21** (2);135-154.

Chapter 5
Fundamental Concepts of Mathematics and Quantum Formalism

Abstract To understand the concept of quantum probability and its application to the cognitive domain, it is necessary to explain the basic concepts of quantum theory. Again, to understand the basic concepts of quantum theory one needs to understand the formalism of Hilbert space. There are several postulates in understanding quantum theory. These postulates are stated in this chapter in a simplistic manner without much mathematical rigor. Von Neumann introduced the projection postulate to understand the measurement process, and this postulate is discussed here in detail. Some experiments like Stern–Gerlach experiment play a crucial role in the development of this theory, a short description of which is given here for convenience. Since mathematical structure like Hilbert space is needed for the mathematical formulation based on these postulates, the various basic notions such as linear vector space, norm, inner product, etc. are defined here. The concept of observable is replaced by the self-adjoint operator in quantum theory. To make grasp those concepts, it is necessary to have some preliminary knowledge about the properties, especially those of self-adjoint operators which are elaborated here. Heisenberg's uncertainty relationships in the context of unsharp observables are discussed. This may help in understanding the current status of research, as well as the developments of the problems related to cognitive science in a more realistic manner.

Keywords Quantum postulates · Projection postulates · Stern–Gerlach experiment · Unsharp observable · Heisenberg uncertainty relation · Vector space · Hilbert space · Cognitive science

In the previous chapter, we discussed the complementary principle, quantum probability and also the double-slit experiment. Almost simultaneously, when, the 'just-now-discussed' developments in various aspects of physics related to various problems of cognitive science had been dominating, two epoch making papers, i.e., of Heisenberg's (in 1969) and Schrödinger's (in 1920) led to the foundation of modern quantum mechanics. Heisenberg proposed the formalism of matrix mechanics, whereas Schrödinger proposed wave mechanics. Afterwards,

© Springer India 2016
S. Roy, *Decision Making and Modelling in Cognitive Science*,
DOI 10.1007/978-81-322-3622-1_5

Schrödinger indicated nearly approximate, physically valid, equivalence between the two approaches. Subsequently, in 1930, Dirac (1982) proposed, though heuristically, a general formalism, embracing both these two approaches. A rigorous mathematical formulation had been introduced, based on the framework of Hilbert space. Now, before applying the axiomatic structure, we need the necessary knowledge about the details of the mathematical tools for formalizing the application of quantum mechanics, including the structure of this approach. Let us discuss some of the necessary postulates as follows:

As such, though, there does not exists any consensus regarding how many axioms are essentially needed in explaining mechanisms dealing with various kinds of problems connected with quantum mechanics, five can be considered as an appropriate number. Among them, the first four postulates can be considered as the basis for the development of the background involved with quantum mechanics, while the fifth one is necessary for supplying the connections between mathematics introduced by the first four and the results of the measurement process. To begin with, let us discuss the necessary postulates connected with these ideas.

5.1 Postulates

Postulate I: A function Ψ is ascribed to define the state of a physical system
Here, the physical system is a microscopic entity like an electron, proton, etc. This function is continuous, single-valued, square-integrable, and globally defined over all space and time associated with each state of such systems. This is a vector in infinite-dimensional Hilbert space. Born (Sakurai 1994) interpreted this function as a probability amplitude. According to this interpretation (Bernstein 2005), the probability of finding a particle is given by the modulus of the square of the amplitude associated with the state function. Born's rule states that the probability of measuring a certain quantum can be defined as a measure of the squared magnitude of its amplitude which, thereby, contains interference terms. But, these terms are absent in classical probability theory. This rule, then, obviously leads to the non-additive nature of the quantum probabilities and, consequently, to the change in states caused by measurements. So, as per Born's rule, the quantum probabilities are decisively dependent on prior measurements, caused by the transitions between two states.

Now, we know that any isolated physical system is considered to be associated with a complex vector space, and the resulting inner product (a Hilbert space) is named as the state space of the system. The state vector, a unit vector in the system, describes completely the physical state of the system. Here, the physical system can be a microscopic entity like an electron, proton, etc. The function thus involved, is continuous, single-valued, square-integrable and globally defined over all space and time associated to each state of such a system. This vector acts as an infinite-dimensional Hilbert space. Thus, regarding the association aspects of a state function with Hilbert space, we can state the following characteristics:

- Hilbert space H of some dimension D associated with every physical system, called the state space of that system, can be completely described by stating its state vector. This vector is a unit vector in that state space. Choosing a particular basis for the Hilbert space, say, $|j\rangle$, $j = 1, ..., D$, a state can be written in the form:

$$|\psi\rangle = \sum_{j=1}^{D} \alpha_j |j\rangle, \quad \text{where} \quad \sum_j |\alpha_j|^2 = 1.$$

- It is possible to write any state in terms of the basis having the spin $= -1/2$, i.e., "spin up" and "spin down", along the Z axis: $|\psi\rangle = \alpha_1 |\uparrow Z\rangle + \alpha_2 |\downarrow Z\rangle$.
- As this has no physical meaning, the overall phase of the state can be stated as $|\psi\rangle$ and $e^{i\theta} |\psi\rangle$ representing the same physical state.
- The choice of a basis can be related to the possible measurements of the system.
- Each basis is considered to be associated with a particular or group of compatible measurement and its outcome.

For the spin 1/2 case, each basis, being associated with a particular direction in space, its component of the spin could be measured along this direction. In such a case, $\{|\uparrow Z\rangle, |\downarrow Z\rangle\}$ denotes the measure of the association for the component of spin along the Z axis, these basis vectors corresponding to spin up or down respectively. In the same way, the bases $\{|\uparrow X\rangle, |\downarrow X\rangle\}$ and $\{|\uparrow Y\rangle, |\downarrow Y\rangle\}$ mean the other possible measurements.

Postulate II

If H_1 and H_2 are two Hilbert spaces associated with the two physical systems S_1 and S_2 respectively, then the tensor product $H_1 \otimes H_2$ of the two Hilbert vector spaces will correspond to the composite system $S_1 + S_2$.

Here, a unitary transformation can describe successfully the time-evolution of a closed quantum system. Then, we can relate the state vector $|\psi\rangle$ of a system at time t_1, to its state vector $|\psi'\rangle$ at time t_2, by a unitary operator U which depends on t_1, t_2, i.e.:

$$|\psi(t_2)\rangle = \hat{U}(t_2, t_1)|\psi(t_1)\rangle = \hat{U}|\psi'\rangle$$

In this case, $\hat{U}$ is independent of the initial state. According to the Schrödinger equation, the equation of time-evolution for the state, then can be written, i.e.:

$$i\tilde{h}\frac{d|\Psi\rangle}{dt} = \hat{H}(t)|\Psi\rangle$$

where $\hat{H}(t)$ denotes a Hermitian operator (the Hamiltonian), describing the energy of the system.

Now the question arises: *How can this be related to unitary transformations?* The answer is that it is quite easy to prove that $\hat{H}$ is a fixed operator (i.e., constant in time), i.e., following a solution to Schrödinger's equation, the evolution of a closed system in continuous time can be written as:

$$|\psi(t_2)\rangle = \exp\left(-i\widehat{\boldsymbol{H}}(t_2 - t_1)/\bar{\boldsymbol{h}}\right)|\psi(t_1)\rangle.$$

Here, the operator $\widehat{\boldsymbol{H}}(t_2 - t_1)$ is Hermitian and, the operator

$$\hat{U}(t_2, t_1) = \exp\left(-i\hat{H}(t_2 - t_1)/\bar{h}\right)$$

is unitary as asserted by $\widehat{\boldsymbol{H}}$. It is a fixed self-adjoint operator known as the Hamiltonian of the system. Then the state at time t is:

$$|\psi\rangle = U|\psi(0)\rangle = \exp\left(-i\widehat{\boldsymbol{H}}(t_2 - t_1)/\bar{\boldsymbol{h}}\right)|\psi(0)\rangle$$

Postulate III: An observable of a physical system can be represented by a self-adjoint operator (or Hermitian) in Hilbert space which allows a complete set of eigenfunctions
The postulate above states that, in Hilbert space, an observable of a physical system is represented by a self-adjoint operator (or Hermitian) which allows a complete set of eigenfunctions.

Now, an observable can be defined as a physical quantity that is measurable by an experiment, and to measure that quantity requires a real number. So the operator corresponding to an observable should be of Hermitian in nature since an Hermitian operator has real eigenvalues. In classical mechanics, an observable is represented by a dynamical variable which obeys Poisson bracket. In quantum mechanics, the Hermitian operators corresponding to observables like position, momentum, angular momentum, etc. are non-commutative in nature. It is important to note that, though classical and quantum mechanics have much dissimilarity, they do share some common features when viewed within a generalized structure (i.e., a Hilbert space). If the state of the system before measurement is taken as $|\psi\rangle$, then the resultant quantum measurement of m occurs with probability $|\psi\rangle$, by operators M. This yields result m with:

$$P(m) = \left\langle\psi\left|M_m^\dagger M_m\right|\psi\right\rangle,$$

and the new state in this case is expressed as:

$$M_m|\psi\rangle/\sqrt{\langle\psi|M_m^\dagger M_m|\psi\rangle}.$$

The probabilities of all outcomes sum to one, so:

$$1 = \sum_m p(m) = \sum_m \left\langle \psi \left| M_m^\dagger M_m \right| \psi \right\rangle.$$

This postulate describes the central point of quantum mechanics: it states that the values of dynamical variables can be quantized (although, in the case of unbound state, it is still possible to have a continuum of eigen-values).

The third postulate, in its second half, describes one of the important characteristics, which can be stated as: after the measurement of $|\psi\rangle$, this yields some eigenvalue $\propto_i$, i.e., the wave function immediately "collapses" into the corresponding eigenstate $|\psi\rangle_i$ (in the case where, $\propto_i$ is degenerate, $|\psi\rangle_i$ become the projection of $|\psi\rangle$ onto the degenerate subspace). Thus, as a result, measurement affects the state of the system. The applications of this finding have extensively been made in many elaborate experimental tests of quantum mechanics.

Postulate IV: Unitary transformation governs the evolution of the state function

The evolution is considered with respect to a continuous parameter, say, time. Here, the unitary transformation is represented by a unitary linear operator U. If we take ψ (t) as the probability amplitude of a quantum state, then the probability amplitude at a later time $t + \Delta t$, will be:

$$|\psi(t + \Delta t)|\psi\rangle = U(t + \Delta t, t)|\psi(t)\rangle$$

Thus, the resultant state space of a composite system can be defined as the tensor product of the state spaces of its component system. All information is stored here, within the state vector, and no passive observation exists, which, essentially, is the dynamics at the microscopic level.

Postulate V: For any measurement process performed, involving an observable, we will get only the eigenvalues of a Hermitian operator corresponding to this observable

Any measurement performed on a quantum state will perturb the state of the system. The measurement instrument is considered to be macroscopic in nature, whereas the system to be observed is a microscopic one. This is in contrast to classical physics, where the instrument, as well as the system to be observed, are both macroscopic in nature.

It is now clear from the above postulates of quantum theory that we need to understand the mathematical structure of Hilbert space so as to understand quantum theory. Now, let us discuss the infinite-dimensional Hilbert space. We start with the properties of vector space on which the inner product is defined, followed by a definition of the metric space.

5.2   Mathematical Preliminaries

5.2.1   Vector Space

Broadly speaking, a vector space can be defined as a collection of objects (set of objects) which behaves like vectors in R^n (on the real-number line in n-dimensions). The objects of such a set are called vectors. Vectors are termed as scalars when multiplied by numbers. Very often, scalars are taken to be real numbers. But, in the case of vector spaces, this may happen with scalar multiplication by complex numbers, by rational numbers or by any field. In 1888, Peano (1888) was the first to formulate the modern, i.e., more abstract treatment, of vector space by encompassing more general objects than Euclidean space. But he treated it as an extension of classical geometric ideas, e.g.: lines; planes; and together with their higher-dimensional analogs. These kinds of applications are, however, mainly applied in mathematics, science and engineering science, etc. They offer a framework for Fourier expansion. With this method, it is possible to apply this framework to image compression routines, as a result of which, also, it is possible to provide an environment in which to employ it in the solution of partial differential equations. Moreover, vector spaces offer an abstract, coordinate-free way of dealing with geometrical and physical objects, for example, tensors. This, in turn, enables examination of the local properties of manifolds with the help of linearization techniques. The vector spaces may also be generalized in many other ways, leading to more advanced notions in geometry and abstract algebra. Mathematically, we can express a vector space as a non-empty set of A objects, called vectors, on which two operations like addition and multiplication by scalars (real numbers) are defined. However, these operations are subject to the ten axioms that must hold for all a, b and c in A and for all scalars, say, α and β. The axioms can be stated as follows:

1. $\mathbf{a} + \mathbf{b}$ is in A.
2. $\mathbf{a} + \mathbf{b} = \mathbf{b} + \mathbf{a}$: commutativity of addition.
3. $(\mathbf{a} + \mathbf{b}) + \mathbf{c} = \mathbf{a} + (\mathbf{b} + \mathbf{c})$: associativity of addition.
4. There is a vector (called the zero vector) 0 in A such that: $\mathbf{a} + 0 = \mathbf{a}$: identity element of addition.
5. For each $\mathbf{a}$ in A, there is a vector (called additive inverse of a), $-\mathbf{a}$ in A which satisfy $\mathbf{a} + (-\mathbf{a}) = 0$, i.e., inverse elements of addition.
6. αa is in A., i.e., identity element of scalar multiplication. This implies multiplicative identity in A.
7. $\alpha (\mathbf{a} + \mathbf{b}) = \alpha\mathbf{a} + \alpha\mathbf{b}$, i.e., distributivity of scalar multiplication with respect to vector addition.
8. $(\alpha + \beta) \mathbf{a} = \alpha\mathbf{a} + \beta\mathbf{a}$., i.e., distributivity of scalar multiplication with respect to field addition.
9. $(\alpha\beta) \mathbf{a} = \alpha (\beta\mathbf{a})$.
10. $1\mathbf{a} = \mathbf{a}$.

The above axioms generalize the properties of vectors, introduced in the above examples. Indeed, the result of addition of two ordered pairs (as in the second example above) does not depend on the order of the summands.

5.2.2 Subspaces

Vector spaces may be formed from other vectors spaces. These are called subspaces. A subspace of a vector space A is a subset H of A with three following properties:

(a) The zero vector of A is in H.
(b) If each **a** and **b** are in H, **a** + **b** is in H. (In this case we say, H is closed under vector addition).
(c) For each **a** in H and each scalar α, α**a** is in H. (In this case, we say H is closed under scalar multiplication).

If the subset H satisfies these three properties, then H is itself a vector. If the scalars are the elements of the field F, we think of a vector space over the field F. In such cases, the vector space is real or a complex vector space, depending on F being real or complete numbers, respectively.

5.2.3 Norms

The concept of norm of a vector is very important. The norm or length of a vector is denoted by $\| \ \|$. This is defined as a function that satisfies the following rules:

1. $\|a\| \geq 0$; $\|a\| = 0$ if and only if $a = 0$
2. $\|\alpha a\| = |\alpha| \, \|a\|$
3. $\|a_1 + a_2\| \leq \|a_1\| + \|a_2\|$ (triangular inequality)

The distance between the two vectors $\mathbf{a}_1$ and $\mathbf{a}_2$ is given by $\|\mathbf{a}_1 - \mathbf{a}_2\|$.

5.2.4 Scalar Product

The scalar product of two vectors can be defined as a scalar-valued function of two vectors. It can be denoted by $f(\mathbf{a}_1, \mathbf{a}_2)$ or usually in the form of $|\mathbf{a}_1, \mathbf{a}_2|$.

If the scalar product exists on a space, the norm also exists and given by:

$$\|\mathbf{a}\| = \sqrt{(\|\mathbf{a}, \mathbf{a}\|)}$$

This follows from Cauchy–Schwartz inequality. The most useful type of basis for the scalar product is an orthonormal one that satisfies:

$$|a_i, a_j| = \delta_{ij}$$

Here, δ_{ij} is Kronecker delta. For $i = j$, this is 1 and zero otherwise.

As the formalism of quantum theory is based on the structure of Hilbert space, now let us describe the Hilbert space on which the formalism of quantum theory is based.

5.3   Hilbert Space

It was David Hilbert (1862–1943) who generalized the notion of Euclidean space (Levitan 2001). This space has typical characteristics, e.g., it is an abstract vector space that has the characteristics of having the structure of an inner product, which makes possible for lengths as well as angles to be measured. Also this space can be stated as **complete**, because it has enough limits within the space to enable the technique to be applied in calculus. Hilbert space can be defined as normed inner-product vector space. Furthermore, this space is complete: there are enough limits in the space to enable the techniques of calculus, also, to be used.

Now, it is to be noted that in discussing the importance of this concept about certain operators, it has been found to play a fundamental role whenever the understanding of the structure of quantum theory is concerned. We know that, in quantum mechanics, the observables are represented by operators only, such as: position operator, momentum operator or energy operator, etc. all known as Hermitian operators. Before going into the characteristics of Hermitian operators, we need to explain some basics about the definition of the operator on Hilbert space, as follows:

Let H be a Hilbert space and the mapping from H into itself is known as operator of H. If M be such an operator, then it will be a bounded operator for any real number α,

$$\text{if:} \quad \|M\| \leq c\|x\| \quad \text{for all } x \in H$$

there exists another operator T^* called **adjoint** of T such that:

$$\|Tx, y\| = \|x, T^*y\|$$

for, all x and y in H.

5.3.1   Hermitian Operator

If the bounded operator T and its adjoint T^* are equal, i.e., $T = T^*$, it is called a Hermitian operator.

Now, the characteristics of a Hermitian operator can be stated as follows:

If T is a Hermitian operator, then

- T has real eigen values.
- An orthonormal basis of eigenvectors of T exists.
- All eigenvalues of T are positive if and only if T is positive and definite.

5.3.2   Unitary Operator

This can be defined as a bijective Hermitian operator whose adjoint is its inverse, i.e.:

$$U\,U^* = I$$

where U is the unitary operator.

5.4   Commutative Properties

In mathematics, the commutative property (or commutative law) can be defined to be a property associated with binary operations and functions. Here, a binary operation on a set is done by calculating the combination of two elements of a set or so-called operands, so as to produce another element of the set. More formally, we can state it as an operation whose two domains together with one co-domain are subsets of the same set. Now, by definition, a binary operation is *commutative* only if, in changing the order of the operands, there is no change in the result. Similarly, when the commutative property holds for a pair of elements under a certain binary operation, only then can it be said that the two elements commute and are said to be under that operation.

It is worth mentioning that the commutative and non-commutative operators are often discussed in the context of quantum formalism. Before going into the details of the commutation and non-commutation of operators, let us define what is meant by commutative and non-commutative for any binary operation say, '$*$' on a set S. It is called commutative:

- if $x * y = y * x$ for all $x, y \in S$, and
- if it is non-commutative, $x * y \neq y * x$ and also,
- If two Hermitian operators T and S commute, we have $TS = ST$.

In quantum mechanics,

- Two Hermitian operators are not, in general, commutative, i.e., $TS \neq ST$, and,
- The non-commutativity of operators is symbolically written as $[T, S] \neq 0$.

Under these conditions, the uncertainty relationship can be derived from non-commuting Hermitian operators T and S so:

$$\Delta T \; \Delta S \geq \frac{i}{2}[T, S]$$

For example, let us start with position and momentum operators in quantum theory that do not commute, i.e.:

$$[X, P] \neq 0.$$

The well-known Heisenberg uncertainty relation between position and momentum can easily be derived from the above operator's inequality as indicated for T and S:

$$\Delta p \; \Delta x \geq \frac{\hbar}{2}$$

where Δp and Δx are variances associated to momentum operator P and position operator X respectively. Here, $\hbar$ is Planck constant h divided by 2π.

The variances or errors in measurements cannot be reduced to zero only for simultaneous measurement of two conjugate variables like position and momentum. This means that it is not possible to measure the position and momentum of any microscopic entity **simultaneously** with infinite precision, i.e., or zero error. There exists another important concept called the projection operator introduced by von Neumann (1932) in connection with the measurement in quantum mechanics.

5.4.1 Projection Operator

Let us consider two linear vector spaces S_1 and S_2 and combine them to construct a new vector space $S = S_1 \oplus S_2$ where the symbol $\oplus$ is called the direct sum. The sum of dimensions of S_1 and S_2 give rise to the dimension of S. The spaces S_1 and S_2 are called subspaces of S. A vector in S can be written as sum of the vectors defined in each subspace. For intuitive understanding of subspaces, let us consider an example in three-dimensional Euclidean space spanned by Cartesian axes x, y, z. Here the x–y plane may be considered as a two-dimensional subspace of the full space and z-axis as a one dimensional subspace. Here, three-dimensional spaces can be thought of as projected on x–y plane, setting the z component as zero.

A projector acts on a vector in full space to make all its components zero excepting the one that is supposed to be projected onto a particular sub-space.

Mathematically, a projection operator P_{S1} on S_1 can be defined as:

$$P_{S1}|\psi\rangle_S = |\psi\rangle_{S1}$$

Equivalently, it can be written as:

$$P_{S1}^2 = P_{S1}$$

where P_{S1} is described as idempotent.

Again, a one-dimensional projector can be written as:

$$P_j = |\phi_j\rangle\langle\phi_j|.$$

Here, two projectors P_1 and P_2 are said to be orthogonal when $P_1 P_2 = 0$. But again, if $P_1 P_2 = 0$, then $P_1 + P_2$ is another projector.

5.5   Projection Postulate (PP)

Von Neumann introduced the projection postulate to understand the measurement process in quantum theory. In the measurement process, the instrument interacts with the system to be observed and the state of the system changes in a way which cannot be described by the Schrödinger equation. In his book Mathematical Foundations of Quantum Mechanics (Sect. 3 in Chap. 3), Von Neumann stressed his view of things, stating: "After the measurement, the state of affairs is ..." and stated that the value measured is one of the eigenvalues of the operator related to it and also opined that the state of the system will be accordingly the corresponding eigenstates. In its simplest way, the postulate states that, when stated in a quite simple form:

If the measurement of a maximal magnitude A with eigenvalues α_1, α_2, α_3 ... and corresponding eigenvectors α_1, α_2, α_3 ... produces the result α_i, then we can state the initial quantum state as the system, transformed into the state α_i. Here, especially, a non-maximal measurement case was considered by von Neumann. Now, if the eigenvalues α_i are assumed to possess multiplicity, namely, k_i, then the corresponding eigenvectors will spread up to a k_i-dimensional subspace, i.e., $k_{\alpha i}$, with the range of a projection operator $P_{\alpha i}$. The conclusion of von Neumann was that the ultimate result of a measurement will be α_i, and the system will be expressed by the statistical operator:

$$P_{\alpha i}/T_r(P_{\alpha i}) = P_{\alpha i}/k_i$$

Here, it should be noted that, according to von Neumann, this should represents a **mixture of states**, but not a pure one. Generally, when an operator represents magnitude A, with a continuous spectrum, following von Neumann's conclusion, we can state that a measurement yields the result, say, $\alpha \in S$, i.e., the system is represented by the unnormalized statistical operator $P_A (S)/P_A (S)$. Here, $P_A (S)$, the projection operator, and A have a spectral measurement with the range $S(A = \int r\, dPA(r))$. In this case, $P_A (S)$ creates the general probabilities. Busch et al. (1995) generalized the concept of the projection operator to understand the physical experiments more realistically, as well as to produce more rigorous formulation of quantum theory. They called the observables unsharp observables.

Before going into the details of unsharp observables, let us extend the concept of adjoint operator T to the unbounded operators on a dense domain D (T). Then an operator is defined as self-adjoint whenever D $(T) = D$ (T^*) and $T = T^*$. A mapping F: B $(\mathbf{R}) \rightarrow L$ (H) is a projection valued measure (PVM), if it satisfies the following:

$$F(X) = F(X)^* = F(X)^2 \quad \text{for all } X \in B(\mathbf{R}), \text{ and, } F(R) = I$$

$$F(U X_i) = \sum F(X_i) \quad \text{for all disjoint sequences } (X_i) \in B(\mathbf{R})$$

B $(\mathbf{R})$ being the Borel subsets on the real-number line, and H is the Hilbert space.

Projection valued measures (PVM) are often used in quantum theory as corresponding to *Sharp Observables*. Positive-operator valued measures (POVM) are introduced by Mittlestaed (2011) by analyzing the physical experiment such as the Stern–Gerlach experiment (1922) from a more realistic perspective, as well as to lay a more rigorous foundation for quantum formalism. Before going into the details of POVM and properties of unsharp observables, we need to understand the Stern–Gerlach experiment from a more realistic stand point, as described in the next chapter.

5.5.1 Statement of Projection Postulate (PP)

Consider a self-adjoint operator associated with an observable o having a discrete, nondegenerate spectrum. Now the measurement of the observable, on a quantum state φ of the system, induces a transition from φ into the corresponding eigenvectors u_k of the adjoint operator $\mathbf{O}$ producing the result a_k.

Historically, two projection postulates have been put forward: the weak projection and the strong projection, as defined in the ensuing paragraphs.

1. **Strong Projection Postulate (SPP)**
   If we select a measurement that leads to particular result a_k, where the state vector $|\varphi\rangle$ undergoes a discontinuous transition, then:

$$|\varphi\rangle = \sum_k c_k |a_k\rangle \rightarrow |a_k\rangle$$

2. **Weak Projection Postulate (WPP)**
   If selections are not made then we consider all possible results of the measurement during the transition, and we get:

$$|\varphi\rangle = \sum_k c_k |a_k\rangle \rightarrow \rho = \sum_k |c_k|^2 |a_k\rangle\langle a_k|$$

Here, we get an ensemble with density ρ.

5.6 Unsharp Observable and Operational Quantum Theory

Sharp and unsharp observables are studied extensively in quantum theory. Sharp observables corresponds to Hermitian operators or their spectral measurement, and unsharp ones correspond to a projection operator known as positive-operator valued measure (POVM). The idea of POVM, dealing with unsharp observables, was introduced by Mittlestaed (2011), trying to resolve the long-standing conceptual and interpretational puzzles in the foundation of quantum theory. It provides powerful mathematical tools to analyze the physical experiments in a more satisfactory and comprehensive way. The physical experiments like Stern–Gerlach experiment play an important role in the development of modern quantum theory. Busch et al. (1995) started by analyzing this experiment, performed in 1922, and introduced this concept, it is generally considered, to prove the existence of the POVM or unsharp observables. We will briefly discuss the arguments of Busch et al. to analyze Stern–Gerlach experiment. One should specify the proper mathematical structures that are relevant to the description of a physical experiment. The Stern–Gerlach experiment has an intrinsic property like spin of the electron. In fact, the first proposal that electrons have spin was made by Uhelnbeck and Goudsmit in (1925). Although the experiment of Stern and Gerlach, though, was originally motivated to test the 'space-quantization' associated with the angular momentum of atomic electrons, it was found later that the interpretation of the experimental results was wrong and the results, in fact, clearly indicate the **existence of spin** of the electron. Let us describe the experiment in the following manner:

5.7 Stern–Gerlach Experiment

In order to test the validity of Bohr-Somerfield hypothesis (Bohr 1958), the Stern–Gerlach experiment was originally intended to prove that the direction of the angular momentum of a silver atom is quantized. Later on, this experiment, proved also the predictions of quantum mechanics for a spin $-\frac{1}{2}$ particle. But, originally, this experiment was seen as a corroboration of the Bohr-Somerfield theory. In the original experiment, a collimated beam of silver atoms produced in a furnace, was made to pass through a inhomogeneous magnetic field, eventually impinging on a glass plate. In the Stern–Gerlach experiment, the spatial trajectory of a particle serves as a detector of its spin state, traversing a magnetic-field gradient. The experiment was run for 8 h, and comparison was made with a similar experiment wherein the magnetic field was eliminated for 4.5 h. When the magnetic field was on, a shape like a pair of lips, in contrast to a single bar, appeared on the glass during the absence of the magnetic field. The maximum gap between the upper and lower lips is approximately on the order of magnitude of the width of the lips. No statistical data was recorded about the distributions at that time, but only visual measurements were made through a microscope. The splitting of the original beam into two distinguishable beams enabled Gerlach and Stern to interpret the findings as proof of space quantization in a magnetic field.

The Stern–Gerlach experiment.

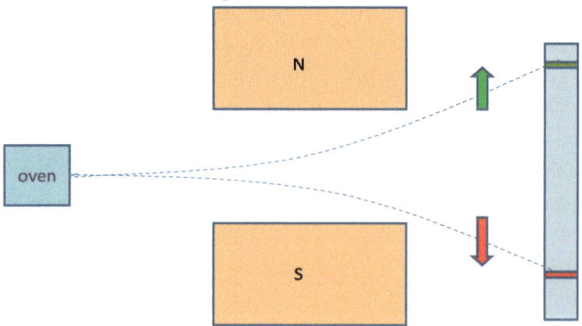

It was realized later that the Stern–Gerlach experiment is an example of a quantum mechanical measurement process which can be used to measure the spin. The beams of the electrons with spin up and that with spin down will be deflected in opposite directions. It is generally believed that the spin of an electron measured with this kind of experiment will be sharp, i.e., either plus one half or minus one half. Busch et al. carefully analyzed this experiment and proved that the possible sources of experimental inaccuracies in this setup are due to the quantum indeterminacy inherent in the center-of-mass wave function. Even if the spin of the electron is prepared with a sharp spin value before the measurement, the value of the spin after the measurement will be uncertain. The unsharpness in the observable is due to various reasons, e.g., due to probabilistic nature of the structure of quantum theory and also to the realistic description of the physical experiment.

5.8   POVM for Spin-Half Particles

A positive-operator valued measure (POVM) can be defined as possessing values, especially on a Hilbert space that is a non-negative self-adjoint operator by nature, and any associated integral must be the identity operator. In dealing with the measurement theory of quantum physics, this approach can be stated as the most general formulation of a measurement. Specifically, the importance of the POVM formalism establishes its importance and necessity, specifically, when dealing with functional analysis and quantum measurement theory. This is so because, whenever it becomes necessary to have the projective measurements on a larger system, we need the required measurements which are by no means to be performed mathematically, by a projection-valued measure (PVM). As a result, when the case of subsystem arises, this will act on that system in ways which PVM is unable to describe alone. In such cases, POVMs are applied, especially in the field of quantum information. Roughly, it can be stated that the POVMs are used on a physical system so that the effect of a projective measurement can be described whenever it is applied on a larger system. In describing the POVM in a simple way, we can define a set of Hermitian-positive semi-indefinite operators $\{F_i\}$ on a Hilbert space, $\mathcal{H}$, where, the sum of the identity operator is:

$$\sum_{i=1}^{n} F_i = I_H.$$

This gives the generalized formula where the decomposition of a (finite-dimensional) space by a set of orthogonal projectors, $\{E_i\}$, are defined, for an orthogonal basis $\{|\phi_i\rangle\}$ by:

$$\sum_{i=1}^{n} E_i = I_H, \quad E_i E_j = \delta_{ij} E_i, \quad E_i = |\phi_i\rangle\langle\phi_i|.$$

In the present case, the main difference is that the elements of a POVM do not require the criteria so that this should be always orthogonal. This characteristic has a great consequence, namely, that the number of elements in the POVM, n, can be larger than the dimension, N, of the Hilbert space, they act in.

Generally, POVMs can also be defined in situations where it is possible for the outcomes of measurements to take values in a non-discrete space which relevant fact indicates that measurement determines a probability measure on the outcome space only.

Definition Let (X, M) be measurable space; that is, M is a σ-algebra of subsets of X. A **POVM** is a function F defined on M whose values are bounded, non-negative, self-adjoint operators, on a Hilbert space H such that $F(X) = I_H$ and for every $\xi \in H$:

$$E \to \langle F(E)\xi|\xi\rangle$$

is a non-negative countable additive measure on the σ-algebra M. This definition should be contrasted with that of the projection-valued measure, which is similar, except that for projection-valued measures, the values of F are required to be projection operators (Wikipedia).

Let $\{\mathbf{n}_\alpha\}$ be a set of unit vectors in R^3 and $\{c_n\}$ be a set of real coefficients such that:

$$\sum c_\alpha \mathbf{n}_\alpha = 0, \quad 0 \leq c_\alpha \leq 1 \text{ and } c_n = 1$$

Then $Q_\alpha = c_\alpha (I + \sigma \cdot \mathbf{n}_\alpha)$, where $Q_\alpha = c_\alpha (I + \sigma \cdot \mathbf{n}_\alpha)$ where σ is Pauli matrix.

This is projection operator for unsharp observables. Now, choosing three vectors $\{\mathbf{n}_\alpha\} = \{\mathbf{n}_a, \mathbf{n}_b, \mathbf{n}_c\}$, we have:

$$c_a = c_b = c_c = \frac{1}{3}, \quad for, \quad n_a + n_b + n_c = 0$$

Then, ultimately, $Q_\alpha = \frac{1}{3}(I + \sigma \cdot n_\alpha)$

For sharp observables, it reduces to the von Neumann projection operator.

References

Bohr, N. (1958) *Atomic Physics & Human Knowledge;* (John Wiley & Sons, New York, 1958), p. 91.

Busch, Grabowski, M. & Lahti, P. (1995); *Operational Quantum theory; Heidelberg: Springer*

Dirac, Paul (1982); *Lectures on Quantum Mechanics*; Oxford University Publications, USA.

Gerlach, W. & Stern, O. (1922); *Der expermentelle Nachweis der Richtungsquantelung im Magnetfeld*; Zeitschrift fur Physik, **9**; 353-355.

Heisenberg, W. (1969) *The physical principles of the quantum theory;* Dover, New York.

Jeremy Bernstein (2005). "Max Born and the Quantum Theory". *American Journal of Physics;* **73** (11); 999–1008.

Levitan, B.M. (2001); *Hilbert space*; in Hazewinkel, Michiel, Encyclopedia Mathematics, Springer.

Mittlestaed, Peter (2011) *Rational Reconstructions of Modern Physics*; Springer.

Neumann, J.Von (1932) *Mathematical Foundations of Quantum Mechanics*; Springer Verlog; *"English translation"*: Princeton University, N.J., 1955.

Peano Giuseppe (1888); Calcolo Geometrico second l'Ausdehnungslehre di H. Grassmann' preceduto dale Operazioni della Logica Deduttiva(in Italian), Turin.

Uhelnbeck, G.E & Goudsmit, S. (1925): Naturwissenschaften **47**, 953.

Sakurai, J. J. (1994) *Modern Quantum Mechanics*; Addison Wesley.

Schrödinger (1920) *Collected papers;* Friedr.Vieweg & Sohn (1984); ISBN 3-7001-0573-8.

Chapter 6
The Complementary Principle, Concept of Filter and Cognition Process

Abstract The complementary principle was introduced by Niels Bohr to explain the mutually exclusive aspects of quantum entities like electron, photon etc. Since Bohr's elaboration, it has been extensively discussed in many branches of science. This principle states that two aspects of a quantum entity, such as wave and particle aspects, cannot be measured simultaneously in a single experiment with infinite precision. Immediately it gives rise to serious ontological issues related to quantum reality. However, developments of understanding in neuroscience lead us to introduce a principle like complementary principle for understanding the functioning of the brain. It is clear from the neurophysiological data that there exist distinct regions responsible for mutually exclusive behaviors such as motor and cognitive, but there may also be other regions that are responsible for both kinds of behaviors. A generalized principle of complementarity has been introduced by the author and his collaborator to describe the response and percept domain using the concept of quantum filters. The quantum filters are generally used for selective measurements. It is shown that the existence of motor and cognitive behaviors can be explained by this principle. This sheds new light on understanding the functioning of the brain and cognitive activities.

Keywords Complementary principle · Percept domain · Response domain · Motor behaviour · Quantum filter · Selective measurement

The motive of this chapter is to show that concepts such as the principle of complementarity, nonlocality, quantum filter, etc., may play an important role in quantum mechanics and also in other branches of science. This extension has been applied as a fundamental structuring principle in constructions of reality by Kafatos and Nadeau (1991). In accordance with these views, we concentrate in this chapter on extending the complementarity principle to neuroscience with the aim of finding the presence of any possible role of this principle in the domain of perception and cognitive aspects (Roy and Kafatos 1999), which has already been established in explaining various problems in quantum mechanics.

© Springer India 2016
S. Roy, *Decision Making and Modelling in Cognitive Science*,
DOI 10.1007/978-81-322-3622-1_6

6.1 Spatiotemporal Representation of Image

In order to recognize a particular object, the brain must go through a matching process to determine which of the many objects it has seen before best matches the object under scrutiny. Moreover, an object may appear in different portions of, in different sizes, and under various orientations on the retina, giving rise to different neural representations at the early stages of the signal system. Research carried out on associate memories gives us some insight into how to handle the problem of pattern matching with the help of neural networks. On the other hand, to produce object representations that are invariant with respect to dramatic fluctuations due to changes in position, size, orientation and other distortions which occur on the sensory inputs, is far from being understood. McCulloch and Pitts (1943) made an early attempt to understand invariant representations from the viewpoint of neurobiology. They hypothesized that the brain forms many geometric transformations of the same sensory input and then averages these using a scanning process (which, they believed, was the role of the alpha rhythm) to form an invariant representation of an object. It can be pointed out that visual attention may provide the key to forming invariant object representations. The main idea is that the process of attending to an object places it into a canonical or object-based reference frame that is suitable for template matching. However, Palmer et al. (1981) made no attempt to describe the neural mechanism for transforming an object representation from one reference frame to another, because theirs was primarily a psychological model.

More recently, Van Essen and Anderson (1989) proposed a neurobiological mechanism for routing retinal information so that an object becomes represented within an object-based reference frame in higher cortical regions. This is known as a 'dynamic routine circuit'. The principal idea behind this approach lies in the fact that it provides a neurobiologically plausible mechanism for shifting and resealing the representation of an object from its retinal reference frame into an object-cantered reference frame. In order to map an arbitrary section of the input onto the output, the neurons in the output stage need to have dynamic access to neurons in the input stage. Here, they proposed that the efficiency of transmission of these pathways is modulated by the activity of central neurons, which have their primary responsibility to dynamically route information in the cortex. Within this framework, they have further developed their ideas and proposed that visual processing at early stages in the cortex is carried out in a relativistic coordinate system, rather than being simply based on absolute retinal coordinates (Van Essen and Anderson 1989). After they analyzed several interesting situations, it was suggested that our perception is based on computations in a relativistic coordinate frame. It was assumed that an image is represented in the spatiotemporal domain and a fixed velocity was considered so as to obtain relativistic coordinates. It should be mentioned that observations carried out on cells in a cat's visual cortex imply the interchangeable role of time and position (Barlow 1979) in determining the apparent position of a stimulus object. Moreover, a certain process of integration must have been followed, because one experiences a single moving object, but not a succession of separate stationary ones. But, when this process of integration is to be followed, in a mathematical sense, it must be carried out in space and time. Not only that, but this process must match that of the moving object, not separately or in the two domains. Space and time are thus seen to play complementary roles in image processing. This is in accordance with the fundamental wholeness of space-time, revealed in its complementary aspects as the unity of space (or "non-locality") and the unity of time as proposed by Kafatos and Nadeau (1991). They assigned complementary status to both spatial and temporal nonlocalities in the sense that, taken together as complementary constructs, they describe the entire physical situation under question, although neither can individually disclose the entire situation in any given instance.

In the cerebellum geometry, we have assumed that a type of nonlocality applies. In other words, the cerebellum geometry is described by a Hilbert space structure. It is worth mentioning that Pribram (1991) constructed a geometry of neurodynamics so that the neural wave functions can be described by a Hilbert space similar to the wave-function in Hilbert space when dealing with ordinary quantum mechanics. He also emphasized two important aspects which should be considered in brain processes, i.e., both a nonlocality similar to the quantum nonlocality and the Hilbert space description for the geometric description should be applied. In this Hilbert space approach, the cortex is assumed to possess a system of eigenvectors forming a complete orthogonal system in the processing sense. Here, we have also assumed a Hilbert space description for the cerebellum geometry where filters for these specific purposes are defined (see below). However, much controversy has arisen during the last decade regarding space-time representation in the brain, particularly for the cerebellum. Pellionisz and Llina (1982) discussed the problems of space-time representation for the cerebellum in a series of publications. In fact, Braitenberg (1967) made a pioneering attempt towards understanding of the space-time representation in the brain. Pellionisz and Llina's critically analyzed the situation and found difficulty in using relativistic coordinates.

In our approach, we have considered specifically a Hilbert space description for quantum filters in the nonrelativistic domain. However, Pellionisz and Llina made some intuitive arguments as to how geometrical structure should be assigned to the cerebellum, where both space and time can be considered simultaneously. At the same time, it is worth mentioning that Pribram considered a relativistic approach to the structure of neural geometry.

6.2 The Response–Percept Domain and Observation Process

As Bohr (1958) and Wheeler (1968) stated: in the world of quantum physics *"no phenomenon is a phenomenon until it is a recorded phenomenon (in this case, image interpretation)"*. In the domain of quantum mechanics, when the experimental apparatus interacts with the observed system, it produces uncontrollable disturbances. In classical physics, one can control these disturbances, in a sense, by modifying the apparatus and repeating the experiments, and it is, in principle, possible to reduce these disturbances to zero. But this is not possible when one has to deal with an experimental situation involving a quantum system. Analogous to a quantum system, image generation always involves the observation of a phenomenon against some background. In other words, the relation of the system-measurement device in quantum mechanics has a formal correspondence to the object-background relation in image interpretation. Again, the background introduces ambiguity with respect to the object similar to the disturbances produced by the apparatus in a quantum-measurement process. An interpretation system can, at a given time, in response terms, only handle a dichotomy relation. It seems to be necessary for anything to be, in some sense computable and interpretable, so that it needs to aggregate certain properties, which, taken together, provide a stable background against which it can be related. This is strikingly similar to the preparation of the state vector in quantum mechanics. What we propose here is a formal similarity between quantum-mechanical measurements and the object background relation in image representation (Roy et al. 1996).

At this stage, everything indicates that the background defines a frame of reference for interpretation. This is equivalent to a restriction of the domain regarding possible responses. Motivated by engineering principles, Paulin (1985) suggested that the cerebellum computes like a Kalman filter. Kalman filtering has been named after Rudolf E. Kalman, who was one of the primary developers of this theory. The theory, also known as linear quadratic estimation (LQE), consists of an algorithm that is applied to those cases where a series of measurements are observed over time. In such cases, the observations thus obtained contain statistical noise as well as other possible kind of inaccuracies, producing estimates of unknown variables. With such advantages, this tends to be more precise than the results obtained from those based on a single measurement alone. Nevertheless, for Kalman filters, it is necessary to have feedback. On the other hand, in modelling cerebellum functions, if we consider its role in cognition, then there should be no feedback control in the percept domain, as the kind that is observed in motor action. So the analogy of using Kalman filters is not very suitable in describing the percept and response domains. It follows that the concept of selective measurement with quantum filters may be more appropriate for the percept–response domains.

First, let us define the state vector, i.e., state of a system at any time t be represented by a normalized state vector $\varphi(t)$ which may be a state for a classical system such as the solar system, pointers or Brownian particles, as well as quantum systems such as electrons and photons. This state vector may also be used to predict or control a process. In that sense, the Kalman-Bucy filter (KBF) can be considered as a state estimator for dynamical systems, because, in the Kalman-Bucy Filter, time taken is considered as continuous, i.e., a counterpart to the discrete time, considered in the Kalman Filter. In any event, similar to the use of the Kalman-Filter, the Kalman-Bucy Filter is also applicable in estimating unmeasured states of a process with the purpose of controlling one or more of them. It was elaborated by Paulin (1985) in his hypothesis, i.e., he proposed that the cerebellum is a neural analog of the KBF. Another kind of filter, i.e., a quantum-state filter (Sakurai 1982) is designed to allow the passage of a particular quantum state and no other. Suppose that the state $|\alpha_j\rangle$ from the eigenbasis $\hat{A}|\alpha_j\rangle = \alpha_j|\alpha_j\rangle$ is incident on a $\hat{B}$-filter designed to allow the state $|\beta_j\rangle$ from the eigenbasis $\hat{B}|\beta_j\rangle = \beta_j|\beta_j\rangle$ to pass. This is equivalent to performing a selective measurement in which only a particular result is acceptable and all others are discarded. We can introduce a similar concept of quantum filters in the response domain. Let us designate them by the term R-filtering. Now the question is: What, in general, can be considered to be the difference between 'classes and symbols' on the one hand and responses on the other hand? It seems that symbols and classes have a greater degree of invariance at the price of less specificity. On the other hand, responses have a greater degree of contextual specificity and consequently less invariance. This greater degree of contextual specificity is closely related to the concept of selective measurements and hence one can introduce the use of quantum filters in the response domain.

At this point, again, when we raise the question of invariance mechanism, for example, during the representation of an object, considering it as a combination of views and the responses involved, it automatically implies, furthermore, a form of equivalence in between structures in the feature domain, as well as in the response domain. Then, we may say that, for the domain of an object, we have a "balance" or equivalence between a particular set of features and a particular response. But, equivalence remains as their combination also manifests invariance. This proves that an entity is not perceivable in the combined percept-response domain interface surrounding this description. This invariance inevitably implies a combinatory balance between the percept and the response domain.

6.3 The Complementarity Principle, Percepts and Concept

The present author along with his collaborator (Roy and Kafatos 1999) discussed the complementary principle in quantum theory and its possible connections with the functioning of cerebellum. It is generally believed that the cerebellum's function is to help the brain to coordinate movements (Holmes 1993). However, the recent neurophysiological evidence challenges this dogma (Barinaga 1996) since the evidence indicates that, apart from being just a specialized control box, the cerebellum participates in many activities of the brain including cognition. It becomes, then, necessary and important to investigate the correspondence between the response and percept domains. Roy and Kafatos (1999) emphasized that, in this case also, there exists a principle similar to the well-known complementarity principle present in quantum mechanics that operates for response and percept domains. The multistable state of perception as well as the various solution properties of response leads us to introduce the idea that quantum filters operate in these domains. Quantum filters are generally used for selective measurements in quantum mechanics. For example, we may consider a Stern–Gerlach arrangement where we let only one of the spin components pass out of the apparatus while completely blocking the other component. More generally, we can imagine a measurement process with a device that selects only one of the eigenstates of the observable and rejects all others. In other way, selective measurement in quantum mechanics only means that this particular eigenstate is termed as *filtration* because only one of these eigenstates filters through the process. Generalizing this concept of filtering, we can think of a filter for selective measurement within the context of the Stern–Gerlach experiment in quantum mechanics. Although filtering measurements are conceivable classically, the main difference between these devices in transmitting wide bands of wavelengths is the fact that classical filter is used in optics and the quantum filter for selective measurements. This can be envisaged by studying what exactly pertains to incompatible observables. However, it must be emphasized that we are taking the idea of a quantum filter at the conceptual level only. Our main objective is to achieve better understanding of the different functions in the brain, e.g., the cerebellum's function. The motive of this discussion is to show the applicability of some concepts of quantum mechanics, such as, the principle of complementarity, nonlocality, quantum filter, etc., which play important roles, besides quantum mechanics, also in other branches of science. In fact, quantum mechanics, in a sense, claims the existence of incompatible observables in micro-physics, which puts impetus on building a unified structure for understanding the conceptual problems related, particularly, to cognition processes.

Bohr (1958), himself, proposed the extension of complementarity principle beyond the atomic realm. Bohr's principle can be summarized as follows:

> …however far the [quantum physical] phenomena transcend the scope of classical physical explanation, the account of all evidence must be expressed in classical terms. The argument is …by the word "experiment" …we refer a situation where we can tell others what we have done and what we have learned,… therefore, the account of the experimental arrangements and of the results of the observations must be expressed in unambiguous language with suitable application of the terminology of classical physics. This crucial point… the impossibility of any sharp separation between the behaviour of atomic objects and the interaction with the measuring instruments…serve to define the conditions under which the phenomena appear….evidence obtained under different experimental conditions cannot be comprehended within a single picture,…must be regarded as complementary in the sense that only the totality of the phenomena exhausts the possible information about the objects (Niels Bohr, "Discussions with Einstein on Epistemological Problems in Atomic Physics" in P. Schilpp; Albert Einstein: Philosopher-Scientist: Open court;1949).

This extension, for example, has been proposed as a fundamental structuring principle in constructing rational complementarity to neuroscience.

The essence of quantum mechanics is that there exist incompatible observables in microphysics. For example, we may consider a Stern–Gerlach arrangement (1922) where we let only one of the spin components pass out of the apparatus while we completely block the other component. More generally, we imagine a measurement process with a device that selects only one of the eigenstates of the observable A and rejects all others. This is what is meant here by selective measurement in quantum mechanics. It is also called *filtration* because only one of the eigenstates filters through the process. Interferometers and diffraction gratings are the instruments to disperse light into its constituent wavelengths. Similarly, filters are commonly used in the domain of optics, specially, in interferometry where these devices are applied to transmit particular bands of wavelength. Various types of filtering are investigated in modern optics (Young 1992).

Thus, generalizing the concept of filter from the domain of optics, we can think of a filter for selective measurement by applying it with the context of Stern–Gerlach experiments as mentioned above. Although filtering measurements are conceivable classically, the main difference between the classical filter used in optics and the quantum filter used for selective measurements can be envisaged by studying what pertains to incompatible observables. In fact, the essence of quantum mechanics is that there exist incompatible observables in microphysics. This concept helps us to build a unified structure for understanding the conceptual problems related to cognition processes. It must be emphasized that in order to better understand the cerebellum function, we are taking here the idea of a quantum filter at the conceptual level. It is important to be mentioned, at present, we are not considering any quantum process that may or not be operating in some regions of the cerebellum, at least at the present state of our understanding of the brain function.

Next, let us start with the case of dentate nucleus present in cerebellum. This is a cluster of neurons, or nerve cells, present in the central nervous system (CNS). The name "dentate" is given due to typical similarity in the structure as it has a dentate —a tooth-like serrated-edge structure—the location lying in the deep white matter of each cerebellar hemisphere. This nucleus, being the largest, lateral and single structure, connects the cerebellum with the rest of the brain. Again, among the four pairs of deep cerebellar nuclei, it is placed farthest from the midline. Parsons and Fox (1999) made a detailed study on the role of the dentate nucleus in the cerebellum using magnetic resonance imaging. The results revealed that the regions of the dentate involved in cognitive processing are *distinct* with respect to their involvement in the control of eye-and-limb movements. So, it may be due to the fact that a kind of quantum filtering is operative in those regions, in the sense, that selective measurements are operating. However, detailed studies of those regions may reveal these phenomena in the near future. The similar kind of selectivity also applies to other regions of dentate nucleus in the cerebellum.

For the present, let us now designate the two filters as R-filters and P-filters, for response and percept domains respectively. An illustration of a quantum-state filter designed to allow only the state $|r_k\rangle$ to pass, is given in Fig. 6.1. Now, suppose that the state $|\alpha_j\rangle$ from the eigen basis $\hat{A}|\alpha_j\rangle = \alpha_j|\alpha_j\rangle$ is incident on the R-filter designed

Fig. 6.1 Illustration of a quantum-state filter designed to allow only the state $|r_k\rangle$ to pass

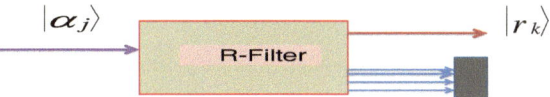

to allow the state $|r_k\rangle$ from the eigenbasis $\hat{R}|r_k\rangle = r_k|r_k\rangle$ to pass. This situation is sketched in Fig. 6.1. The probability that the state $|r_k\rangle$ is obtained after passage through the filter, given an incident state $|\alpha_j\rangle$, is determined from the usual quantum expansion coefficients, to be $\langle\alpha_j|r_k\rangle^2$. In other words, we have the usual expression:

$$|\alpha_j| = \sum_k |r_k\rangle\langle r_k|\alpha_j\rangle$$

But, during the measurement of the property represented by $|\alpha_j\rangle$, we need to follow the statement of quantum theory which states that the probability of obtaining the eigenvalue r_k is the square modulus of the overlap $r_k|\alpha_k\rangle$. After the measurement, the state is projected onto the corresponding eigenstate of $|\alpha_j\rangle$. A sequence of two quantum-state filters designed to allow only the state $|r_k\rangle$ of $|\alpha_j\rangle$ to pass, is illustrated in Fig. 6.2.

However, we can equally envisage the expansion of $|\alpha_j\rangle$ in the eigenbasis of the $\hat{P}$-filter as well so that:

$$|\alpha_j\rangle = \sum_m |p_m\rangle\langle p_m \mid \alpha_j\rangle \tag{6.1}$$

The probability of obtaining $|r_k\rangle$, given an incident $|\alpha_j\rangle$, can now be written as:

$$\begin{aligned}|\langle\alpha_j \mid r_k\rangle|^2 &= \left|\sum_m \langle\alpha_j \mid p_m\rangle\langle p_m \mid r_k\rangle\right|^2 \\ &= \sum_m\sum_{m'} \langle\alpha_j \mid p_m\rangle\langle p_{m'} \mid r_k\rangle\end{aligned} \tag{6.2}$$

We can think of $|\alpha_j\rangle$ as being "made up" of the states $|p_{m'}\rangle$ and the probability that we are ultimately interested in $|\langle\alpha_j|r_k\rangle|^2$, can be viewed as resulting from an interference of all the possible paths of the kind, namely:

$$|\alpha_j\rangle \rightarrow |p_m\rangle \rightarrow |r_m\rangle \tag{6.3}$$

Let us consider another situation, where a P-filter is actually inserted before the R-filter, as shown in Fig. 6.2. If the P-filter is set to admit the state $|p_{m'}\rangle$, then the probability P of obtaining $|r_k\rangle$, given an input $|\alpha_j\rangle$, is now given by:

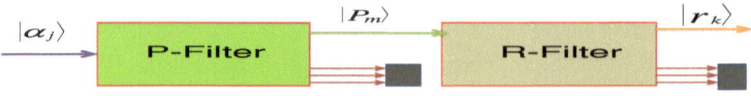

Fig. 6.2 A sequence of two quantum-state filters designed to allow only the state $|r_k\rangle$ to pass

$$P = \left|\langle \alpha_j \mid p_m \rangle\right|^2 \left|\langle p_m \mid r_k \rangle\right|^2 \tag{6.4}$$

Let us sum over all the possible filter settings to obtain the total probability as:

$$P = \sum_m \langle \alpha_j \mid p_m \rangle \langle p_m \mid \alpha_j \rangle \langle r_k \mid p_m \rangle \langle p_m \mid r_k \rangle \tag{6.5}$$

But this is not equal to Eq. (6.2) due to the existence of cross terms in it. These terms are called interference terms. In fact, we can write this equation as:

$$\left|\langle \alpha_j \mid r_k \rangle\right|^2 = P + \sum_m \sum_{m'} \langle \alpha_j \mid p_m \rangle \langle p_m \mid r_k \rangle \langle r_k \mid p_{m'} \rangle \langle p_{m'} \mid \alpha_j \rangle \tag{6.6}$$

where the second term specifies the term containing the contribution from interference, having all possible $|\{p_m\}\rangle$ paths. The $\hat{P}$-filter, in the two-slit experiment, is equivalent to determining the precise trajectory of the particle and Eq. (6.5) is a mathematical expression of the consequent destruction of the interference. Now, in the context of percept and response domains, we can regard $|p_m\rangle$ as all possible alternatives that cause the interference, but, the interference will ultimately be destroyed when passing through the response domain.

For example, when we recognize an object such as a line or an edge, we get a localized object. This implies that the interference-like terms should vanish to allow these localized objects to be obtained.

Now, looking at Eq. (6.5), what we get is the interference term to vanish if the second term goes to zero, i.e.:

$$\langle \alpha_j \mid p_m \rangle \langle p_m \mid r_k \rangle \langle r_k \mid p_{m''} \rangle \langle p_{m'} \mid \alpha_j \rangle = \delta_{mm'} \tag{6.7}$$

If $|p_m\rangle$ becomes simultaneous, the eigenstate of both $\hat{P}$ and $\hat{A}$ (and/or $\hat{R}$), only then will this condition be satisfied. For this to have occurred, P and A (and/or R) must be compatible observables. In another way, we can write:

$$[\hat{A}, \hat{P}] = 0 \quad \text{and/or} \quad [\hat{R}, \hat{P}] = 0 \tag{6.8}$$

The action of the $\hat{P}$-filter can be detected if $\hat{P}$ represents a complementary property to both of the properties. In other words, a particular system state with a given response is equivalent or complementary to a particular percept. That means, the preceding must be true for all the interfaces between levels where invariants are formed, implying that this must be valid for all levels of a system. Moreover, the implications should be true also for the interface of the entire system, together with its environment. Not only that, but, if this is to be viewed from the wider percept-response domain, these implications are also valid for the combination of percepts that a system experiences, the responses it produces and they constitute equivalence. Finally, it depicts that, for the entire system as well, they appear as an invariant to the environment in a wider domain. Usually a system can be externally observed from its response only, and we cannot expect the effects in this subdomain invariant, or otherwise the system could not affect its environment. Then, the interesting question immediately could be raised: *What happens when there is no response motor action even when there is activity in the percept domain?*

In such a case, the phase difference between the state vectors in the percept and the response domains may not be 90°, the interference term will not be zero, and we may have partial interference. It appears from the above analysis that there is a possibility to describe the percept and response domains using the concepts of state vectors in the Hilbert space and the idea of quantum filters due to the presence of an interference term between the vectors, defined in the percept and response domains. Interestingly, this appears analogous to the double-slit experiment in quantum mechanics. The percept and response domains are said to be equivalent if the interference term vanishes. Thus it becomes the similar situation while detecting the complementary aspects such as those of the wave and particle in the double-slit experiment.

Let us try to explain this phenomenon in the following way: in the percept domain, we can recognize an object such as a line, edge, etc. This appears to be a kind of localization phenomenon analogous to the particle aspect. But in the response domain, we have a kind of signal only, and a process of non-localization begins. So this becomes similar to the wave aspect, and then these two domains seem to be complementary to each other. Indeed there might be a correspondence between the percept domain and the response domain. We can think of relativistic transformations such as the Lorentz transformations between the different frames of reference in these two systems. This has been emphasized, also, by Van Essen (1989). We can think of a rest-frame in the percept domain and the motion or activity in the response domain. So there may be a kind of time dilation that could, in principle, be calculated using the Lorentz transformation. This time dilation causes the phase difference between the state vectors, defined on the percept and response domains. Then, depending on this phase difference, eventually, the interference term for the filters in the percept domain and the response domain will be destroyed.

An interesting situation arises when there is no motor action, but it has been observed that, simultaneously, something exists there in the cognition process that has been detected in recent neurobiological experiments (Parsons and Fox 1999). It seems that there is no time dilation between the frame of reference for the percept domain and the response domain. If we consider the frame of reference attached to the percept domain, the phase difference between the two vectors in the two domains now will never be $\Phi/2$ or $3\Phi/2$ so as to get a vanishing interference term. The vectors may be in phase lag of 0° or 180°. Here we have a situation analogous to the retaining of the interference term in quantum processes, so no detection of events is possible. But if there is a partial interference (for phase angles less than $\Phi/2$), the detection of different complementary aspects is possible, similarly, for the detection of a hazy path with partial interference.

Usually it is not possible to measure simultaneously the two complementary aspects with arbitrary accuracy. But one can accomplish this with certain unsharpness introduced in the measurement of the complementary aspects. Then the observables are known as "unsharp" observables. This kind of unsharpness may arise due to various physical situations. For example, as stated earlier, in the case of measurement of the spin of an electron, we use a magnetic field in the Stern–Gerlach type of experiment, but, due to various physical factors such as the presence of thermal noise, etc., we do not, in actuality, measure the exact value of spin as, e.g., $+1/2$ or $-1/2$. What we observe is a distribution for the spins. It is proposed here that it is necessary to analyze the details of neurophysiological experiments to gain more insight into the physical causes required to reduce the observables into unsharp states. As a result, there will be an overlapping of complementary features. In this case, mutual exclusiveness expressed by the term complementarity refers to the possibilities of predicting outcomes, as well as specific value determinations. Bohr (1948) and Pauli (1994) discussed both these aspects.

In a recent review, Busch et al. (1995) analyzed this situation in detail. They considered two types of complementarity: one is referred to as the measurement-theoretical complementarity and the other as the probabilistic complementarity. The former implies non-coexistence. In the case of sharp observables, the two formulations are equivalent. However, for unsharp observables, in the measurement, theoretical notion of complementarity is stronger than the probabilistic one. This is due to the fact that, in the framework of probabilistic complementarity, the simultaneous measurements of complementary observables like position and momentum, or path and interference, are possible. So we can construct a joint observable for a coexistent pair of unsharp path and unsharp interference observables. Bohr's original concept of complementarity, referred to as strict mutual exclusiveness, is thus confirmed for pairs of sharp observables. On the other hand, Einstein's attempt to evade the complementary verdict can also be carried out in some sense. But for this, the price has to be paid by introducing a certain degree of unsharpness in measuring the complementary aspects. In this chapter, we have discussed a generalized principle of complementarity that is applicable to structures beyond the quantum realm, similar to the cases found in the case of the percept and response domains. Here, we have discussed the problems related to complementarity in neuroscience which has already been discussed in one of monographs by Kafatos and Kafatou (1991).

In modern cognitive neuroscience, the prevailing concept is that cognitive functions are active predominantly at the network level. Actions of individual neurons might be limited only in forming the simple basic elements of these networks. This conceptual framework envisages that, in many individuals, perhaps, there exists a favorable configuration of the micro-architecture within the cognitive-implicated network. As a result of these characteristics, they might exhibit outstanding cognitive abilities. Not only that, but their final formation in ontogenesis may occur in a relatively random way (Atmanspacher and Römer 2012). Also, the recent neurophysiological results (Middleton and Strick 1997) suggest the projection of cerebella output itself via the thalamus into multiple cortical areas. This output includes not only the premotor and prefrontal cortex, but also the primary motor cortex. According to their views, the projections to different cortical areas may originate from distinct regions of the deep cerebellar nuclei. As mentioned earlier, Arshavsky (2003) expresses the view that the cognitive functions are performed basically at the network level. He also opined that the specific function meant for individual neurons does not appear to extend beyond the formation of simple basic elements of these networks. But, some neurobiologists (Bloedel and Bracha 1997) have challenged this kind of distinct behavior of the cerebellum (Middleton and Strick 1997) after studying some patients, though they have acknowledged that cognitive and motor functions are integral constituents of the mechanism governing animal behavior. They analyzed substantial amounts of data supporting their view that, if the cerebellum is important for executing a specific behavior, it also participates in other kinds of longer- or shorter-term modifications of the characteristics of the behavior.

It is now clear that there are many ambiguities regarding the various experimental results. The problem arises regarding the integration process while coordinating dentate regions, responsible for motor and cognitive actions. Thus, the generalized complementarity principle as proposed by Roy and Kafatos (1997) might play the role of coordinator among this kind of diverse results. It is clear from the neurophysiological data that there exist distinct regions responsible for mutually exclusive behaviors such as motor and cognitive, but there may also exist other regions [as pointed out by Bloedel and Bracha (1997)], which are responsible for both kind of behaviors. Roy and Kafatos (1997) showed in their framework that for unsharp observables and P- and R-filter algebra, there are distinct regions indicating the existence of incompatible observables. These other regions, as indicated by Bloedel and Bracha, (1997), indicate the possibility of joint measurements of incompatible observables.

These observations might be a clear indication of the probabilistic complementarity as discussed in this chapter. As described earlier, here, the selective measurements involved in the percept and response domains have a strong similarity with the concept of quantum filters used in quantum mechanics. Some regions of dentate nucleus in the cerebellum may be responsible for this kind of selective measurement. But the problem arises due to the presence of the integration process, while coordinating other dentate regions of the cerebellum, that is mainly responsible for motor actions. In this case, more refined experiments in neurophysiology should be performed so as to shed new light on our understanding of the cognition process and responses. The complexity involved in the human brain has not yet been subjected to the level of detailed mathematical description required for full scientific knowledge. It can be emphasized that, even if specific quantum-mechanical processes are not directly involved in explaining the cognition process, quantum-like concepts such as complementarity, quantum filters, and nonlocality may play significant roles in a generalized approach. As such, complementarity may extend beyond the atomic realm and may be a powerful means to explain phenomena such as those involved in cognition, as envisaged by Bohr.

Finally there is another important aspect worth mentioning with regard to the use of the concept of quantum filters for selective measurements. Normally, in describing motor action, we need to predict a kind of displacement, i.e., some hand or limb movement or some eye movement. Here, the measurement is reduced to measuring displacements. Also, in most of the experiments in quantum physics, a measurement normally refers to measuring displacement such as the movement of a needle. But in the cognition domain, however, we do not know exactly what type of measurement is being performed. We do not even know whether we can at all describe those events in terms of physical measurements. But, interestingly in the percept domain, we may think of a physical measurement as being neurophysiological, similar to the absorption process in some of the quantum experiments. More detail neurophysiological experiments are required to clarify these issues.

References

Arshavsky, Yuri, I. (2003) Brain Research Reviews; 41(2-3), 229-267.

Atmanspacher, H., & Römer, H. (2012) Journal of Mathematical Psychology; **56**, 274-280.

Barlow, H.B. (1979); Nature **279**, 189.

Barinaga, M. (1996); Science; **272**; 482.

Bloedel, J.R. & Bracha, V. (1997); *The Cerebellum and Cognition* ed. by J.D. Schmahrnann (Academic Press, San Diego); p. 613.

Bohr, Niels (1948) *On the notions of Causality and Complementarity*; Dialectica; Wiley Online Library.

Bohr, Niels (1958) *Atomic Physics & Human Knowledge;* (John Wiley & Sons, New York,), p. 91.

Busch, P., Grabowski, M. & Lahti, P. (1995) *Operational Quantum Physics* (Berlin: Springer Verlag), ISBN: 3-540-59358-6.

Braitenberg, V. (1967) Prog. Brain Res; **25; 334.

Holmes, G. (1993); Brain; **62**; 1.

Kafatos, M. & Nadeau, R. (1991) *The Conscious Universe;* Springer- Verlag, New York.

Kafatos, M. & Kafatou Thalia (1991) *Looking in and seeing out; consciousness and cosmos*; Quest books; First edition.

McCulloch, W.S. (1943) & Pitts, W. (1943); Bull. Math. Biophys.; **5**;115.

Middleton, F.A. & Strick, P.L. (1997) *The Cerebellum and Cognition,* ed. by J.D. Schmahrnann; Academic Press, San Diego; 61.

Parsons, L.M. & Fox, P.T. (1999); PET & FMRI studies of Cerebellar Function in Sensation, Perception and Cognition, in *PET: Critical Assessment of Recent Trends,* eds. by Muller-Gartner and B. Gulyas (Kluwer Academic Publishers, Dordrecht, Netherlands).

Palmer, S., Rosch, E. & Chase, P. (1981); Attention and Performance; **9**, 135.

Pauli, Wolfgang (1994) *Writings on Physics and Philosophy*; Springer.

Paulin, M.G. (1985) *Cerebellar Control of Vestibular Reflexes,* Ph.D. Thesis, University of Auckland.

Pellionisz, A. & Llinas, R. (1982); Neuroscience; **7**; 2949.

Pribram, K. (1991) *Brain and Perception-Holonomy & Structure in Figural Processing; Lawrence Eribaum Associates Publishers; Hillsdale, NJ.

Roy, S., Kundu, M.K. & Granlund, G.H. (1996); Inform. Sciences. **89**; 193.

Roy, S & Kafatos, M. (1999) Phys. Essays; 12(4); 662-668.

Sakurai, J.J. (1982) *Modern Quantum Mechanics;* Addison-Wesley, New York.

Van Essen, D.C. & Anderson, C.H. (1989); in *Introduction to Neural and Electronic Networks;* eds by J.L. Davis & Lau, C. (Academic Press, Orlando).

Wheeler, J.A. (1968) *Super space and the nature of quantum geometrodynamics,* in Batelles Rencontres (B. S. De-Witt and J. A. Wheeler, eds.), Benjamin, New York, 1968.

Young, M. (1992) *Optics and Lasers;* 4th ed. (Springer-Verlag, New York).

Chapter 7
Quantum Probability Theory and Non-Boolean Logic

The only acceptable point of view appears to be the one that recognizes both sides of reality—the quantitative and the qualitative, the physical and the psychical—as compatible with each other and can embrace them simultaneously.
—Pauli (1955: 208).

Abstract Since the very inception of quantum theory, the corresponding logic for quantum entities has attracted much attention. The logic underlying the quantum theory is shown to be non-Boolean in character. Boolean logic is a two/valued logic which is used for the description of everyday objects. Modern computers are based on this logic. The existence of an interference term for microscopic entities or quantum entities clearly indicates the existence of three-valued or non-Boolean logic. This is popularly known as quantum logic. It is mathematically shown that a set of propositions which satisfies the different axiomatic structures for the non-Boolean logic generates Hilbert space structures. The quantum probability associated with this type of quantum logic can be applied to decision-making problems in the cognitive domain. It is to be noted that, until now, no quantum mechanical framework is taken as a valid description of the anatomical structures and functions of the brain. This framework of quantum probability is very abstract and devoid of any material content. So it can be applied to any branch of knowledge like biology, social science, etc. Of course, it is necessary to understand the issue of contextualization, i.e., here, in the case of the brain.

Keywords Non-Boolean logic · Quantum logic · Quantum probability · Hilbert space · Contextualization

In this chapter, we will try to describe non-Boolean logic for a broad audience consisting of logicians, physicists and other branches of researchers in cognitive science, including philosophers of science. These communities come from diverse backgrounds, so far as their research domains are concerned. For example, normally a physicist does not have formal training in logic, and a cognitive scientist is by no means expected to master Hilbert space formalism in quantum mechanics. Thus, it

© Springer India 2016   131
S. Roy, *Decision Making and Modelling in Cognitive Science*,
DOI 10.1007/978-81-322-3622-1_7

turns out to be a real challenge to write a chapter on quantum probability and non-Boolean logic for such a broad community. Nowadays, by the category of logic, we mean a formal logic where logical systems are based on a precisely defined formal language. In this case, logical consequence is considered as a key concept in logic. Briefly speaking, logical consequence is a relation between two statements or between two sets of statements. In the beginning, let us start with an introduction to the concepts of logic related to the essentials in dealing with cognition and decision making processes. Then we will try to deal with the concept of Boolean logic, as well as its contributions and controversies in those branches, in a more general base.

7.1   Logic and Cognition

McNamara (1994) discussed logic and cognition in a comprehensive manner. The two subjects, i.e., logic and cognition, are either remotely connected or not connected at all. Logic dealing with the problem of specifying a correct inference is dealt with in the domain of philosophy, whereas the other one, i.e., cognition is dealt with in the domain of psychology, because it is usually viewed as the mental activity. Now the serious debate lies in finding the relationship between logic and psychology. One group maintains that the foundations of logic have a psychological basis. Friedrich (1773–1843) and Mill (1806–1873) belonged to this school of thought. On the contrary, Husserel and Moran (2001) in his book "*Logical investigations*" argued that:

(a) Logic does not derive its basic principles from psychology.
(b) Logic does not describe psychological states and events. And then, Hacck (1978) took the third position, i.e.:
(c) Logic has nothing to do with mental processes.

It is interesting to note that it was McNamara who proposed that logic and psychology constrain each other like mathematics and physics constrain each other. For example, calculus was discovered as a mathematical language for analysis of the continuum, which has been used to understand the Newtonian law of gravitation. Similarly, logic is considered as a mathematical language to understand the cognition processes. Works on decision making consider logic to be an integral part of understanding the decision-making process.

7.2   Logic and Decision Making

Logic is generally considered as the philosophy of valid reasoning. It is primarily studied in disciplines such as philosophy, mathematics, semantics and computer science. George Boole (1815–64), an English mathematician, developed a system of logical thought known as Boolean logic. "AND, OR and NOT" are common Boolean operations used in this logical system. Usually, truth tables and Boolean algebra are used to describe logical expressions. Thus, formal logic can, in general, be defined as the principle of identity. Boolean logic is extensively used in computer science. "True or false" are two values which can be assumed by a variable within the Boolean framework. The applicability of this algebra is effective, especially, in that particular formal machinery where it is allowed to logically combine this kind of truth values. Thus the mechanism of Boolean operations offer a framework by which it is possible to handle the questionnaire and also the resulting data in such a way that this can be handled by the common binary kit ("Yes" or "No"). In various decision-making processes, e.g., clinical decisions, even measurements and probabilities, it is needed to have the data, very often, that is *dichotomized* (*binaries*). Many common examples can be found in many devices, e.g., digital computers and on-or-off electronic switches like, which are connected in a Boolean manner. However, the framework of Boolean operations and also their logical combinations underlie the logical checking of '*rule-based, decision-support systems*', for instance, for inconsistencies, incompleteness and redundancy.

7.3   Boolean Algebra

Let X, Y, Z be the Boolean variables which hold the answers to 'did test X come out positive?' Then,

- $\neg X$ = (not X) = (false if X is true; true if X is false) = (**F** if X, otherwise **T**), where, **T** and **F** denotes the two truth values True or False, respectively.
- It is important to note that $\neg(\neg X) = X$; Other basic operations are "AND and OR":
  AND,
- ('Did both X and Y comes out positive?'): $X \wedge Y$ = (X and Y) = (**T** if both X and Y, otherwise **F**).
  OR,
- ('Did X or Y, or both, come out positive?'): $X \vee Y$ = (X or Y) = (**F** if neither X nor Y, otherwise **T**) = $\neg(\neg X \wedge \neg Y)$.

The mirror image of the right-most identity also works in the following way:

$$X \wedge Y = \neg(\neg X \vee \neg Y) = (\textbf{F} \text{ if one or both of } X \text{ and } Y \text{ are false, otherwise } \textbf{T}).$$

In set theory, intersection ($\cap$), union ($\cup$) and complementing are the precise analogs of $\wedge$, $\vee$ and $\neg$, respectively. OR and AND are *associative* operations: more than two terms can be OR (AND), in arbitrary order, to reflect 'at least one true' ('all true'). Next, *Distributive* properties include:

$$(X \wedge Y) \vee (X \wedge Z) = X \wedge (Y \wedge Z).$$

$$(X \vee Y) \wedge (X \vee Z) = X \vee (Y \wedge Z).$$

and

$$(X \wedge X) = (X \vee X) = X.$$

Finally, test X must be either positive or negative, but cannot be both:

$$(X \vee \neg X) = \mathbf{T}, \quad (X \wedge \neg X) = \mathbf{F}.$$

But, the former expression is a *tautology*, i.e., a necessarily true proposition.

In case of decision making for robots, there does not appear any ambiguity. The choices made by them should be very clear. As a result, their decisions are always based on the answers to questions that have only two possibilities: "true or false"; "yes or no", i.e., binary logic is followed.

The developments of modern physics, especially the quantum physics, question the validity of Boolean logic. The computers are thought of as the study of constructing a "universe" while physics tries to understand the universe in which we find ourselves, i.e., so called "physical reality". The values of logic (true or false) correspond to objects or states in the physical world, which we construct by observation only. These values are discrete, created by measurement. In quantum theory, if the superposition of states collapses into one of its state, then only it corresponds to real valued logic. Otherwise, the existence of interference phenomena, which appear in quantum theory for microscopic entity, like electron and photons, prohibits the existence of logical values like "yes" or "no" only. This indicates that classical Boolean type of logic is not valid in the framework of quantum theory. Both logic and emotion play important roles, acting almost simultaneously, in case of decision making. Based on logic as well as on the senses, it is necessary to figure out the best options among multiple ideas, and then take decision about what is logical. But the limitless capacity for processing innumerable information does not exist in the case of a decision maker. For humans, a bound on the limits of emotions and cognitive capacities is always there whenever the question of applying their abilities regarding those two capacities arise. Specially, these two criteria, among other possible aspects, drive human beings in engaging themselves, employing some rational thinking and assessing the possible consequences before finalizing any strategic step about making an ultimate decision.

A smart decision or the best result depends on the accumulated data, past experiences and dynamic variables, together with their combinations. Thus, whenever we make a decision following our logical thinking, we try to exclude our emotions and, instead, we stress our logical, rational methods of thinking together with some mathematical tools, if possible. In fact, the foundation of such logical decision making lies in the principle of utility, so that the value of each option can be assessed by putting some weighted criteria on each consideration. Simon (1972), in his famous work of 1957, proposed the idea of "bounded rationality', in which he raised reservations about making perfectly rational decisions because of the presence of finite capacity in the information processing of human beings. He pointed out that humans are not only partly rational, they are, most of the time, emotional and little bit of irrational in certain aspects of their final actions. So, whenever their information processing is taxed with multiple tasks, bounded within their cognitive capacities, people try to process information sequentially, say, applying some heuristics such as rules of thumb. Simon suggested that this mode, following the above techniques, improves the capacity of the performance of human and so, acts as a limited information processor. He argues: this process *"is at the heart of all human intelligence"*.

But, Payne et al. (1992) looked at the problem from a different angle and commented that humans act as adaptive decision makers, always trying to assess "the benefits" of a decision strategy, applying the strategy of maximum probability, so that, they achieve the best possible results as an alternative against, specifically, time, cost, mental efforts, money needed, etc. Buying a car could be a good example of that. Hence, we can explain this evaluation as a part of the information process which uses the non-compensatory decision rules to control multi-attribute decision problems. Later, Amos Tversky and Daniel Kahneman (2003) discussed various aspects of human judgment, emphasizing the heuristic-biases approach as the cause for the biases and errors, very much unlike the previous approach where the use of heuristics is considered as trade-offs between efforts and accuracy. They opined that it was simply a failure to recognize the "correct" solution, and they considered these heuristics as cases which, possibly often, might be highly effective and economically the most viable. But, at the same time, they also pointed out to the possibilities that could happen in some cases which might lead to systematic and predictable errors. They did Putnam's example in this context: Sometimes, the objects or events are judged as frequent, probable or causally efficacious, but, the extent of this kind of actions might be dependent on their ready availability in memory.

Recently, in judgment and decision research, much attention has also been paid to the emotional side. It has been noticed that, when cognitive evaluations become responsible for emotional responses to risky situations (say, worry, fear), the responses have greater impacts on risk taking behavior than the cognitive aspects of evaluations. Take the case of rash driving where some drivers may underestimate positive-valued risks. But, in the case of airplane travel, some people are extremely fearful thinking of flying, even while recognizing the risks to be low, and so the negative-valued activities may lead to overestimation of risks. Thus, depending upon the degree of intensity of emotions, this can totally overwhelm deliberative decision making. Also, the indirect influence of immediate emotions could appear in cases where people's judgments regarding their expected consequences are involved. A quite common example for such a case: People, who are not hungry or not in pain, under-appreciate the intensity of feelings of those who are hungry or in pain. Furthermore, the effects of immediate emotions cannot be neglected when interpreting information. The decision makers, in such cases, attend selectively to, and retrieve, emotionally biased information. Extensive studies have concluded that "*negative emotions narrow attentional focus. In contrast, positive emotions broaden attentional focus*". So, the emotion-based decision can be stated as a little different from that based on rational logic. In such cases, dynamic variables are also considered together with data collection, because the whole range of decision making uses emotion, which depends on the degree of logic included with this process. Typically, a completely emotional decision is quite fast because it takes a very short time (~ 0.1 s) for the rational cortex to get going. This process can be termed as reactive or subconscious decision-making, which could very often be found when en-countering heated arguments or facing immediate danger. However, in the case of common emotional decision-making, some logical arguments are present. But, the emotion being the driving force, there always exists a two-way possibility, i.e., either using a pseudo-logic to support emotional choices (extremely common) or just overriding the logical arguments involved.

As explained earlier, another kind of decision making is not rare, i.e., when the use of emotion starts with logical arguments but terminates as totally emotional in the final choice. But, in logical decision making, the decision depends on past experiences and how one feels about these past experiences. In all such cases, we can find many such empirical findings in decision making, especially in the domain of cognition, which clearly *shows the existence of a term like interference* in estimating the sum of the probabilities of two mutually exclusive events. The existence of such interference-like terms assigns the logical values, not only **Yes, or No,** but also an **intermediate** value which clearly proves the inadequacy of Boolean logic in the cognitive domain.

7.4   Quantum Logic and Non-Boolean Algebra

In 1936, John von Neumann and famous mathematician Garret Birkhoff jointly published a paper entitled "The logic of Quantum logic". We can comfortably state that the following conclusion is responsible for the start of the idea and development of quantum logic. They concluded in their famous paper of 1936 (considered as the most quoted paper (BvN) in quantum logic): "*Hence we conclude that the propositional calculus of quantum mechanics has the same structure as an abstract projective geometry.*" (Birkhoff and von Neumann, emphasis in the original 1936).

This conclusion had a great impact on existing conservative ideas and opened up the new vista in developing the algebraic logic, as well as non-classical algebraic structures that possess much weaker properties than Boolean algebras. Many efforts have been made to understand this non-Boolean structure in the context of quantum theory since its very inception. Birkhoff and von-Neumann, in their classic paper, first tried to elaborate the concept of quantum logic. They made several attempts to reanalyze the conceptual structure, and thereby a new logical model, that applies the Hilbert space formalism of quantum mechanics. This formalism is known now as *quantum logic*. As their paper was not easily readable and needed to be interpreted and reconstructed in terms of modern terminology, in spring 2005, Dove Gabby,

Kurt Engesser, Daniel Lehmann and Jane Spur put forward an excellent idea–to ask the experts of quantum physics and quantum structures to contribute–to the *"Handbook of Quantum Logic and Quantum Structures"*.

In essence, the quantum logic can be defined as a set of rules that deals with some reasoning and propositions. But, for this, it is essential to consider the necessary principles of quantum theory while developing the theory behind it. Certain properties connected to quantum logic have been found distinctively different than those of classical logic. For that, it is possible to proffer an example, i.e., the failure of the distributive law of proposition logic. As per definition, quantum logic can be defined in the terms of lattice and Hilbert space, i.e., "it is a Hilbert lattice". This set of closed subspaces consists of an infinite-dimensional complex Hilbert space. It is important to note that these ortho-modular lattices are frequently studied equations since their applications are very important for the various problems of perception and decision making.

In fact, an easy and comfortable way of introducing the concept of quantum logic is to introduce this logic in terms of "experimental propositions' that are able to define certain non-classical connectiveness, explicitly, for a certain special class of propositions. This class of propositions can then be subjected to the idealized quantum mechanical tests. However, these cases will, arguably, be suited to the limited subject matter under considerations. In this way, one could achieve the expected goal by introducing a local non-classical logical system that satisfies certain essential requirements. Putnam (1974) made this argument and claimed it to be essential for accepting different logics in order to adopt and connect to different subject matters. However, the explanation can be viewed from the angle of sketching the quantum logic as descriptive and to connect this to the empirical behavior of certain experiments (especially in the cognitive domain). It is known that one can approach quantum logic through the axiomatic and abstract approach, based on "yes-no" test, known as the Geneva school (Jauch and Piron 1969). Moreover, it has also been established by Bell (1986), who first introduced a framework by employing the notion of attributes over a space with a distinguished lattice of subsets and then showed that this logic reveals its differences from classical physics in a quite intuitive way, i.e., this can be established intuitively through the fact that quantum logic makes room for the characteristic quantum-mechanical notions (of course, through properly formulated versions) of superposition and incompatibility of attributes. So far as the meaning of attributes is concerned, we can state them as the qualities such as like 'hardness', 'blackness' or having a particular charge, such as positive, being possessed by or manifested over a space (maybe called a manifold or field). Considering, for example, the field to be a sensory field, the part of this field manifesting blackness and, say, the part depicting hardness, ultimately describes a blackboard manifesting both these attributes. Here, each attribute correlates itself with a proposition (or, precisely a propositional function).

7.4.1   *Propositional Logic*

Propositional logic is by definition is not concerned with the structure of propositions beyond a certain point where it is impossible to be decomposed anymore with the help of logical connectiveness. But the inferences can be restated by replacing those atomic statements with connected statement letters. This, in turn, can be expressed as variables representing those statements. The language of propositional logic is built up in modern logic by using the following symbols:

- A set of propositional variables
- Symbols of the connectives: negation ¬ and implication →
- Brackets

7.4.2   *Lattices*

Mathematics defines lattices which can be characterized as algebraic structures, satisfying certain axiomatic identities. Due to this equivalence, the lattice theory is extensively used both in order theory and universal algebra. There are semi-lattices, which include Heyting and Boolean algebra, and hence, all these "lattice-like' structures admit both order-theoretic as well as algebraic descriptions. Let us begin with the definition of the concept of lattices. Following Engesser et al. (2011), we write:

Definition (I) A partially ordered set is a pair $\langle \mathbf{L}, \leq \rangle$, where $\mathbf{L}$ is a non-empty set and $\leq$ is a binary relation satisfying the following relations:

1. $A \leq A$ for any $A \in \mathbf{L}$ (reflexivity)
2. If $A \leq B$, and $B \leq A$ then $A = B$ (anti symmetric)
3. If $A \leq B$ and $B \leq C$ then $A \leq C$ (transitivity)

where, $C \leq$ as partial order.

Definition (II) The partially ordered set $\langle \mathbf{L}, \leq \rangle$ is called a lattice if, for any two elements A and B, there exists the least upper bound $A \vee B$ (logical 'or' or disjunction), the greatest lower bound $A \wedge B$ (logical 'and' conjunction) and negation (logical "not' or ¬). Also, there exists a zero element and a unit element 1. The lattice is called complete for any subset of $\mathbf{L}$ having a greatest lower bound but a smallest upper bound.

Definition (III) A lattice is distributive if the following condition holds for any $A, B, C \in \mathbf{L}$, i.e.:

$$(A \vee B) \wedge C = (A \wedge B) \vee (A \wedge C)$$

Definition (IV) Let **L** be a lattice. Then a map $A \rightarrow A^{\perp}$ is called an orthocomplement and, $A^{\perp}$ the orthocomplement of A. If, it has the following properties, i.e.:

- $A^{\perp\perp} = A$
- If $A \leq B$ then $B^{\perp} \leq A^{\perp}$

We call $\langle \mathbf{L}, \leq, ^{\perp} \rangle$ an orthocomplimented lattice if $\perp$ is an orthocomplement of the lattice $\langle \leq \rangle$.

Definition (V) A Boolean algebra is an orthocomplemented and distributive lattice. Let us now, discuss ***Quantum logic using some of tools introduced in propositional logic***:

Classical probability theory and its various modifications are based on Boolean algebra over W. We define a set F of subsets of W that contains W which is closed under union, intersection and complementation. Here, the elements of F may be events or propositions. In quantum probability, the events or propositions are *vector subspaces,* and the mathematics of quantum theory is based on vector spaces, i.e., Hilbert spaces (1953). The algebra of these subspaces is non-Boolean with respect to the operations, i.e., intersections, unions and ortho-complementations. The set of all such subspaces forms an ortholattice L (H) (H is Hilbert space) with $p \leq q$; where, p is a subspace of q. Here, logic operations, i.e., "and", "or" and "not" are represented by $\wedge$ (infimum), $\vee$ (supremum) and $(^{\perp})$ (orthocomplement) on L (H). With the help of this kind of elementary knowledge, it is possible now to construct the formal language of quantum logic.

Let the proposition p semantically entail another one q. Now, whenever the propositions are not members of a common Boolean sublattice, i.e., they are incompatible, quantum logic is non-distributive. Under such condition, let p, q, r be incompatible propositions, then $p \wedge (q \vee r)$ and $(p \wedge q) \vee (p \wedge r)$ may denote distinct subspaces and, so, are not semantically equivalent.

We will discuss this non-Boolean logic structure of quantum logic and quantum ontology in the next chapter.

References

Aristotle (born 384 BCE, Stagira, Chalcidice, Greece—died322, Chalcis, Euboea); http://www. britannica.com/EBchecked/topic/34560/Aristotle

Birkhoff Garret; Von Neumann, John (1936) Annals of Mathematics; **37** (4): 823–843.

Bell, J.L. (1986) Brit. J. Phil.Sci.; **37**, 83-99.

Engesser, K, Gabbay, D.M. & Lehman, D. (2011); "*Handbook of Quantum Logic and Quantum structures: Quantum Structures*"; Eds. by Kurt Engesser, Dov M. Gabbay, Daniel Lehmann; Elsevier, Radarweg 29, PO Box 211,1000 AE Amsterdam, the Netherlands.

Friedrich, Jacob Fries (1773-1843); Chisholm, Hugh, ed. (1911). Fries, Jacob Friedrich; *Encyclopædia Britannica*; (11th ed.); Cambridge University Press.

Hacck, Susan(1978) *Deviant Logic, Fuzzy Logic: Beyond the Formalism*;The University of Chicago.

Hilbert David & Corant, Richard (1953); *"Methods of Mathematical Physics"*;Vol I, Interscience.

Husserel Edmond, Moran Dermut (2001) *Logical Investigations, Vol.1*, abridged; reprint; revised; Psychology Press.

Jauch, J.M. & Piron, C. (1969) On the structure of quantal Proposition systems; Helvetica Physica Acta; **36**, 827-837.

Mill, John Stuart (1806-1873) *Bulletin of Symbolic logic; 19th century Logic Between Philosophy and Mathematics*; **6**(04); Dec 1999; 433-450;

Mill, J. Stuart (1806-1873) *On Liberty and other Writings*; (Extends the 1974 *Deviant Logic*, with some additional essays published between 1973 and 1980, particularly on fuzzy logic, cf. *The Philosophical Review*, **107**: 3, 468–471.

Payne, J.W., Bettman, J.R. & Johnson, E.J. (1992) *Behavioral decision research: A constructive processing perspective*; Annual Review of Psychology, **43**, 87-131.

Putnam, H. (1974) How to think quantum-logically"; Synthese, **29**, 55-61. Reprinted in P. Suppes (ed.); *Logic and Probability in Quantum Mechanics* (Dordrecht: Reidel, 1976); pp. 47-53.

Simon Herbert (1957); 1957. *"Models of Man"*; John Wiley; ibid; (1972); (with Allen Newell): *"Human Problem Solving"*; Prentice Hall, Englewood Cliffs, NJ, (1972).

Simon Herbert, A. (1972); *"Decision and Organization"*; by C.M. Mcguire & Ray Radner (eds.); North Holland Publishing company

Chapter 8
Quantum Ontology and Context Dependence

> *The only acceptable point of view appears to be the one that recognizes both sides of reality—the quantitative and the qualitative, the physical and the psychical—as compatible with each other and can embrace them simultaneously.*
>
> —Pauli (1955, p. 208).

Abstract Recent advances in understanding quantum reality lead to the proposal of quantum ontology. Here, as such, there is no distinction between the classical and the non-classical world. This is based on the abstract framework of propositional calculus which gives rise to Hilbert space structure, in which case, the framework is devoid of any material content like the concept of elementary particles and their localizations. The fundamental constants such as the Planck constant (h), speed of light (c) and gravitational constant (G), which have definite numerical values, need to be interpreted in this abstract framework. This is known as contextualization in the arena of modern physics. Some attempts have been made by Mittlestaed and his collaborators (Mittlestaedt et al. 2011) in this direction. They tried to understand this type of contextualization based on the idea of POVM (positive-operator valued measure) and unsharp observables. The Planck constant has been shown to be the degree of unsharpness in the observability of complementary variables like position and momentum in the context of the Heisenberg uncertainty principle. To apply the concepts of quantum ontology and quantum probability in other branches of knowledge such as the cognitive domain, it is necessary to make a prescription for contextualization.

Keywords Quantum ontology · Fundamental constants · Contextuality · Unsharp observables

In his seminal paper, "Two dogmas of Empiricism (1951)", the logician and philosopher W.V. Quinne argued that all beliefs, in principle, should be subject to revision, if different kinds of empirical data, including the so-called analytic propositions, are taken into consideration. Thus the laws of logic, being paradigmatic cases of analytic propositions, should not be taken as immune to revision. So,

© Springer India 2016
S. Roy, *Decision Making and Modelling in Cognitive Science*,
DOI 10.1007/978-81-322-3622-1_8

to justify this claim, the so-called paradoxes of quantum mechanics have been revealed. Birkhopf and von Neumann (1936) tried to resolve those paradoxes. They proposed that, if the principle of distributivity is abandoned, then it would be possible to substitute quantum logic for classical logic. Although, Quinne (1951) did not provide any sustained argument for the claim in that paper, he also did not seriously pursue the argument. But, in *Philosophy of Logic* (the chapter titled "Deviant Logics"), again this idea was rejected by Quinne when he commented that classical logic should be revised with respect to the response posed by the paradoxes. The argument behind his concern was that it will cost "a serious loss of simplicity" and "the handicap of having to think within a deviant logic" (Quinne 1976). However, he stood by his comment and claimed that logic is 'in principle' not immune to revision.

As a next point, the question may arise: Is this theory is a counterintuitive "measurement problem"? It is well known that the complete description of a quantum system is given by a wave function defined as a vector in Hilbert space (an abstract vector space). This statement becomes difficult to understand if the related problem is derived from a realistic point of view. In classical mechanics, in Maxwell's theory of electromagnetism, as well as in the general theory of relativity, waves act on the particles and guide their motion. In this kind of case, the term ontology is used for something which exists independent of human observation or even independent of the existence of the human race, but whose evolution is governed by laws of physics. Now, let us start with the question: *"What is really going on in nature when we are not giving any attention to it or not at all interested to think it is necessary to know or need to have prior knowledge about it?"*

Only when we require definite answers to this kind of inquiry, i.e., a reality outside of thinking or perceiving the very subject, does the objective reality arise. Whenever we are asked how, and in what sense, this so called anything may exist when we are not looking at it or independent of our interest, only then are we making an ontological inquiry. This means we are getting the act of reality of being as such, without considering the attributes of this or the entity itself. Following the same argument, when questioned about the reality of wave functions, characterizing every possible kind of physical entity, we are really asking the question of ontology.

In the quantum paradigm, it is very hard to introduce such a concept called "quantum ontology". The wave function, an abstract vector, enters into an algorithm. It predicts accurately the "results of measurement". However, Griffiths (2009, 2011) tried to construct a consistent quantum ontology based on two main ideas: (i) The logic, constructed here, is shown to be compatible with the Hilbert-space structure of quantum mechanics, and (ii) The quantum evolution is considered to be an inherently stochastic process, not just only when measurement is taking place. The main advantage of this framework is that it has no measurement problem. The consistent quantum ontology essentially resembles the ontology of classical mechanics represented by a deterministic Hamiltonian. Recently, Mittlestaedt et al. (2011) discussed the existence of possible quantum ontology where no material content of the microscopic domain is used, rather, language and logic is used to describe microscopic entities. In this way, both classical ontology (i.e., the ontology of classical mechanics) and quantum ontology are investigated within this abstract framework. Since this framework is devoid of any material content, it can also be applied to any branch of knowledge.

The framework based on quantum logic is based on an abstract framework, completely devoid of material content and, hence, can be applied not only to physics but also to fields such as biology and social sciences. But, the most challenging issue remains embodied in the question: how to apply it to a particular branch of knowledge. In modern physics, fundamental constants like the Planck constant (h), the speed of light (c) and the gravitational constant (G) play very important roles, and one should take care of these constants within the above abstract framework of quantum ontology. Mittlestaedt (2011) discussed this problem within the framework of operational quantum theory, based on the concept of unsharp observables. This showed the way in which the quantum ontological framework can be made context dependent, so as to include the characteristics of the microscopic domain. In principle, it can be applied to the domain of cognitive science. However, the ultimate aim, here, is to study the workability of the above framework, connected to context dependence and also applied in the cognitive domain.

To start with, let quantum mechanics be considered as universally valid in contrast to the Copenhagen interpretation. Then we could proceed in the query whether classical physics is really intuitive and plausible. These problems are discussed within the quantum logic approach to quantum mechanics. There, the classical ontology is relaxed by reducing metaphysical hypotheses. Now, on the basis of this weak ontology, we will try to explain a formal logic of quantum physics. Mittlesteadt (2002) primarily dealt with this problem in detail by introducing an orthogonal lattice and established this logic. However, by means of the Solar's condition and Piron's results (Piron 1976), one can reach the same conclusion considering classical Hilbert spaces. Mittlesteadt (2011) replaced this ontology with that of unsharp properties and concluded that quantum mechanics is more intuitive than classical mechanics and that classical mechanics cannot be taken as the macroscopic limit of quantum mechanics.

It is now well established that descriptions of quantum-mechanical events are basically context-dependent descriptions. The role of quantum (non-distributive) logic is in the partial ordering of contexts, rather than in the ordering of quantum-mechanical events. Moreover, the kind of quantum logic displayed by quantum mechanics can easily be inferred from the general notion of contextuality, as used in ordinary language. Heelan (1970) discussed this in detail and showed that the formalizable core of Bohr's notion of complementarity is the type of context dependence which can be applied, in principle, to the domain of cognitive science. However, the issue is to make the above framework context dependent for the cognitive domain. We briefly discuss in this chapter, for convenience, how the framework of quantum probability has been made context dependent, even in modern physics itself. Before going into the issue of context dependence, let us start with the very concept of quantum ontology.

8.1 Newton and Metaphysics

It is often argued that Newtonian mechanics is based on many hypotheses that can be traced back to their origins in seventtenth century metaphysics and theology. Isaac Newton primarily based his physics on absolute time and space. But, he also followed and adhered to the principle of relativity of Galileo Galilei. However, this was hidden and never made explicit. The success of the Newtonian framework in understanding the phenomena of nature did not raise questions regarding the metaphysical aspects of the above-mentioned hypotheses. Many of the scientists (Giuliani 2001; Lange 1885), and especially Mach (1883), invoked the argument of absolute time and space and stated them as "essentially as metaphysical concepts, and should not be considered as scientifically meaningful". He suggested the relative motion between material bodies to be a useful concept in physics. Following Newton, he also argued that the effects depend on accelerated motion with respect to absolute space and cited the example of rotation in order to show that it can be described purely, with reference to material bodies. Also, he insisted that it might be possible to relate the inertial effects, which Newton cited in support of absolute space, purely, to acceleration with respect to a fixed star. For example, Newton in his "principia" wrote: "*Absolute true and mathematical time, of itself and from its own nature, flows equably without relation to anything external*".

However, there is no evidence of rational or empirical justification for this statement in Newton's writings:

> In recent literature, Newton's theses regarding the ontology of space and time have come to be called substantialism in contrast to relationism. It should be emphasized, though, that Newton did not regard space and time as genuine substances (as are, paradigmatically, bodies and minds), but rather as real entities with their own manner of existence. (From: Plato.stanford.edu/)

Mach (1883, 1912), Poincaré (1889, 1901) exposed the metaphysical aspects of the Newtonian framework by the end of nineteenth century, and, in particular, it was done by Einstein in the beginning of twentieth century. Henri Poincare´ proposed, in his work related to the "relativity principle", as a general law of nature (including electrodynamics and gravitation), but only in 1905 did he use this principle while correcting Lorentz's preliminary transformation formulas. In his pioneering work, Poincare´ presented an exact set of equations, called the Lorentz transformations (Cushing 1967). Almost at the same time, Albert Einstein published his original paper "Special Theory of Relativity", where he independently derived the Lorentz transformations, based on the relativity principle. This has been radically reinterpreted as changing the fundamental definitions of space and time intervals. With this approach, he avoided the need for any reference to a luminiferous ether as stated in classical electrodynamics, abandoning the absolute simultaneity of Galilean kinematics. The idea of ether had been proposed in the nineteenth century, in Lorentz's ether theory (LET), which he formulated in the period 1892–1895), considering the wave theory of light as a disturbance of a "light medium" or luminoferous ether.

8.2 Quantum Ontology

When we ask a question like "what is going on in nature when we are not looking at it?", in a real sense, we believe in objective reality, i.e., a reality outside of thinking about or perceiving a subject. But whenever we want to know, in what sense, anything may exist, even in the absence of any kind of observation, we are already stumbling on an ontological inquiry. This is because, in such a situation, we are not dependent on any attributes of this or that entity, but absolutely on the 'act' or 'reality of being' as such. Or, stated in another way, we can define "ontology" by referring it to the philosophical tradition that claims the existence and ability to recognize an independent world are outside the existence of things in themselves ('Dinge ansich', in the sense of Kant)—like the "real" people whose shadows are perceived by the prisoners. Following the same argument, whenever we ask, for instance, a broad question about the reality of wave functions or states characterizing each and every possible type of physical entity, we are stumbling on, in a true sense, asking about the ontological aspects of it. So far as the development of modern physics is concerned, when the modern physics approaches the limit of what can be physically observed, modern physicists, at a certain of stage, directly engage themselves with the development of the various metaphysical questions behind their already developed theories. This happened, especially, during the several stages of development of quantum mechanics which deals with the various types of crucial questions of ontology, i.e., the science of being.

We can summarize in brief, the most widely known approach in interpreting quantum mechanics, i.e., the "Copenhagen interpretation", basically, a collection of varying opinions proposed by Niels Bohr (1925–1927) and Werner Heisenberg, among others. Common features of these proposals can be listed in short as:

1. The wave function as a probabilistic description of phenomena;
2. Bohr's "complementarity principle, where matter exists simultaneously, as a wave and a particle";
3. It is impossible to know non-commuting properties simultaneously,, following the Heisenberg principle of uncertainty.
4. Matter may exist simultaneously in two contrary, well-defined states, following the principle of superposition.
5. Reality is altered due to the act of observation, and the underlying reason may be the possible existence of the "collapse of the wave function" and other possible paradoxes.

At this point, some physicists started looking at all these quantum paradoxes as a result of the practical limitations of measurement apparatus and also the processes involved with this. On the contrary, mathematical formalism for the development of quantum mechanics, unlike its philosophical interpretations, has been experimentally verified as an excellent device for predicting the probability distributions of events. As mentioned earlier, Maria Solér (Aerts and Steirteghem 1999) developed the mathematical formalism of quantum mechanics and discovered from his analysis this formalism which allows, for a purely lattice, the theoretical characterization of classical Hilbert spaces. He introduced the ontology of unsharp properties and concluded that quantum mechanics is more intuitive than classical mechanics. It is important to note here that quantum mechanics can be obtained from classical mechanics merely by reducing the ontological premises. It is not needed to incorporate new empirical components which, in fact, can be easily be explained within the framework of the quantum-logic approach to quantum mechanics. It was Mittlesteadt, who, in his famous work (Mittlesteadt 2003), showed that there is no plausible justification of Solér's law (1995) and that the quantum ontology is nothing but partly too weak and partly too strong. As stated by Mittlesteadt:

> The ontology of a certain domain of physics contains the most general features of the external reality which is treated in the physical domain in question. In particular, the ontology should contain the material preconditions for a pragmatics which allows for the constitution of a scientific language and thus for the formulation of physical experiences.

We can define classical ontology when that ontology underlies classical mechanics. Following Mittlesteadt, let it be denoted by O(C). It has been pointed out, and discussed by him in a profound manner, how it could be possible to raise important objections against O(C) (Mittlesteadt 2011). But, it has now been already established that the metaphysical and theological reasons of Newton no longer remain relevant for the justification of the ontology. Then, it becomes necessary to search for alternative reasons that might be the possible cause for its drawbacks. In fact, we may opine from our basic knowledge that some of our daily experiences does not tally with some aspects, i.e., the strict causality law, the unrestricted conservation of substance and also the existence of one universal time. Due to these reasons, we cannot consider these and other hypotheses of the classical ontology (O(C)) as intuitive and plausible.

Mittlesteadt also raised arguments against the experimental evidence presented by the mentioned hypothesis indicating the experimental suggestion that objects can always be individualized and re-identified at later times. It can be stated that the principle of complete determination lacks experimental proof possessing that grade of accuracy which, in any way, could claim the result as an established principle. Also, it is not possible to find any justification for the absence of a strict causality law. If it would have been present, the present state of an object would allow for predictions about all elementary properties, which principle also fails to find any empirical justification for O(C). The origin for the base of these hypotheses can be traced back to the metaphysics of the seventeenth and eighteenth century. Hence, O(C) appears neither to be neither intuitive nor plausible, and, not being justified by experimental evidence, it cannot be stated as in accordance with quantum physics. It was Mittlesteadt who formulated with success the quantum ontology as a reduced version of the O(C), by relaxing and weakening some hypothetical requirements of O(C). It is important to note that no new requirements need to be added to the assumptions of the classical ontology O(C).

We will briefly characterize this ontology and analyze this, based on classical or Newtonian mechanics as follows:

- *O(C)* [1]: *There exists an absolute time. It establishes a universal order of two or more events; it provides a universal measure of time and explains the concept of simultaneity of two, spatially separated events.*
- *O(C)* [2]: *There exists an absolute space. It explains the concepts of absolute motion and absolute rest. Euclidean geometry applies to this absolute space.*
- *O(C)* [3]: *There are individual and distinguishable objects. These objects can not only be named and identified at a certain instant of time, but also re-identified at any later time.*
- *O(C)*[4]: *These objects possess elementary properties P_λ in the following sense: An elementary property P_λ refers to an object system such that either P_λ or the counter property $\overline{P_\lambda}$ pertains to the system. Furthermore, objects are subject to the law of thoroughgoing determination according to which "if all predictions are taken together with their contradictory properties, then one of each pair of contradictory opposites must belong to it".*
- *O(C)*[5]: *for objects of the external objective reality, the causality law holds without any restriction. There is an unbroken causality.*
- *O(C)* [6]: *for objects of the external objective reality, the law of conservation of substance holds without any restriction.*

The ontology based on these requirements is called classical ontology O(C). These requirements are neither justified by experimental evidence nor by intuitive understanding. In modern physics, especially after the birth of the special theory of relativity and quantum theory, a new ontology has been proposed by relaxing some of the requirements in O(C). It took a few steps to reach what we call now quantum ontology as O(Q). The first one is known as O(SR), i.e., ontology based on the special theory of relativity (SR), and then O(GR), i.e., ontology reconstructed based on the general theory of relativity (GR). By relaxing the requirements of absolute time and absolute space (done within the framework of special relativity), one obtains the reduced ontology O(SR). This describes a wider domain of reality than that conceived in classical ontology O(C). Mittlestaed proved how the finite and definite value of the speed of light can be understood within this reconstructed ontology O(SR), which does not depend on definite law or particular methods of observation. Quantum ontology O(Q) can be reconstructed from classical ontology O(C) (Mittelstaedt 2011) in the following manner:

- *If an elementary property P pertains to an object property, then a test of this property by measurement will lead with certainty to the result P.*

- *Any elementary property P can be tested at a given object with the result that either P or the counter property $\overline{P}$ pertains to the system.*
- *Quantum objects are not thoroughgoingly determined. They possess only a few elementary properties, either affirmative or negative. Properties, which pertain simultaneously to an object, are called "objective" and "mutually commensurable".*

It is evident from the above lists of requirements that the first two requirements exist also in classical ontology O(C), where one gets the third one by relaxing requirement in O(C). However, the more general quantum ontology has been reconstructed based on the idea of unsharp observable and POVM as discussed in the previous chapter. We designate this ontology as $O(Q^U)$. To consider a wider domain of reality so as to reconstruct $O(Q^U)$, one needs to relax the following requirements in O(Q):

- *The most general observables in quantum mechanics correspond to unsharp properties that allow for joint properties, even for complementary observables. Hence, O(Q) is too restrictive since it does not allow generally for joint properties.*
- *The requirement of value definiteness cannot be fulfilled for all properties, since the pointer-objectification in the measurement-process cannot be achieved generally. Hence, O(Q) is also not sufficiently restrictive.*

It has been established by Mittlestaedt that, interestingly, starting from quantum ontology and then following a long sequence of arguments, we can reach quantum logic, akin to an orthonormal lattice and then through it, step by step, ultimately to the classical Hilbert space, i.e., from quantum ontology to the classical ontology (see Mittlesteadt 2011 for details). However, this abstract theory even now is unable to contain Planck's constant h. Mittlesteadt, in his famous works on quantum ontology, explained this aspect as follows: "*h can be obtained, if the empty theory is applied to real entities*".

But for this to occur, the concepts should be extended to that extent, so that classical notions will be considered in usual cases. But, in order to define the objects, with the help of a symmetry group as well as systems of imprimitivity, we need to introduce the concepts of localizability and homogeneity. It is well known that, for an elementary system, the irreducible representations of the Galelio groups are projective, which can be determined only up to a parameter z ($z = m/h$, $m =$ mass of the particle and $h =$ Planck's constant). It can be shown that h has a meaning within the quantum mechanics, even if we use the classical concepts in the relevant derivation.

In this general quantum ontology $O(Q^U)$, the Planck constant has not been addressed so far. However, the Planck constant has a definite numerical value. On the other hand, within the framework of $O(Q^U)$, there does not exist any border between the classical and quantum worlds. So, it is not necessary to consider the Planck constant in the formulation of O(Q) in a Hilbert space framework. Within the framework of $O(Q^U)$, the Planck constant is meaningful in the following sense

(Mittlestaedt 2011: 98): "*In the spirit of uncertainty relation, we can say that the minimal degree of unsharpness of probabilistically complementary properties which pertains jointly to a system, is given by $\bar{h}$. This is the meaning of Planck's constant on the level of quantum ontology.*".... "*The largest possible degree of joint determination of unsharp complementary properties*".

As a conclusion, Mittlesteadt (2011) finally admits that the quantum logic of unsharp propositions is not the "true logic". Since the unsharp propositions are unable to solve the measurement problem, this logic does not claim this proposition as the "final logic" of physics which could claim itself as a universal "final theory of everything". But, it can be stated that the unsharp quantum logic is "closer" to the "final logic" (quoting Mittlesteadt) than other previous proposed logics, i.e., orthomodular logic or classical logic. However, there exists substantial criticism in the scientific community of physicists and philosophers against the "logic of quantum mechanics". There, they have raised their objections stating: This is mechanics, not a "genuine logic" in the strict sense, following the formal structure in the tradition of Aristotle, Thomas Aquinas (Kenny 1976) or George Boole (1854) by which our rational thinking and arguing are governed. Finally, regarding different views of quantum logic, the following conclusion can be drawn with Mittlestaedt's statement:

> Since quantum mechanics is usually considered as an empirical structure that corresponds to a genuine law of nature, also the quantum logic, which finally leads to quantum mechanics must contain empirical components that at the end of this way of reasoning, imply the empirical components in quantum mechanics (Rational Reconstructions of Modern Physics; Mittleseadt 2011, springer, p 73).

In the present context, the question may arise whether the laws of quantum logic are any longer laws of nature. The answer to this statement could be formulated in the way that it is a formal structure which evolves a priori from the corresponding preconditions of a scientific language of quantum physics. Moreover, it could be posed that the laws of quantum logic are the laws of nature. This logic is quite strongly dependent on the concerned context only. To express it in a straightforward way, it is context dependent. Not only that, but:

> it has been well observed conclusion that quantum mechanics is at the bottom an empty theory, a formal framework that must be filled step by step with empirical content. .... Quantum ontology is only a framework that must be filled with linguistic and mathematical structures. .....Quantum mechanics is based on a weaker ontology than classical physics and depends, for this reason, on less, not sufficiently justified hypothesis... theory valid in the large domain of reality that corresponds to the weak ontology. (Mittlesteadt 2011: 75)

So far as the context dependence of modern physics has been discussed, it is clear from the above discussions that this is a very abstract framework devoid of any material content, for instance, the concept of elementary particles like the electron and proton or the quark or Higgs boson, etc. It can be applied to any branch of knowledge like social science, biology and so on. Many factual phenomena have been observed in many cases where apparently there exist many phenomena that break the classical probability laws, but, the systems related to such phenomena are context dependent and hence adaptive to other systems. The famous double-slit experiment is the well-known example of context dependence. Similar kinds of problems can also be found both in biology and psychology. Context dependence is very much present, and so it is a problem active, especially in cognitive science, where the influence of environmental factors plays a vital role on one's perception about a stimulus.

It is already an established fact that states of biological systems can well be represented with vectors in linear space, observables, and by operators, whereas corresponding quantum mechanics applications provide the necessary probabilistic interpretations. In order to represent the information about behavior and dynamics of biological systems, quantum mechanical aspects, i.e., the irreducible contextuality of these systems, play a crucially important role. It becomes necessary to make comparisons between biological and quantum contextualities (especially, adaptivity to the contexts, such as the environment). Generally, adaptivity, in quantum mechanics is expressed with respect to open systems and corresponding generalizations. Recently, Asano et al. (2013) applied non-Kolmogorovian approach to study the context-dependent systems for which the violations of classical probability law had been observed. According to their views, the irreducible contextuality of biosystems is the main source of quantum representation of information about their behavior and dynamics. They showed that the contextuality and adaptivity of biosystems leads to violation of the laws of classical probability theory (based on the Kolmogorov axiomatic of 1933). It has been noticed that these features of quantum probability, having a non-Kolomogorovian character, can be used to analyze some phenomena which are connected to statistical properties, but are quite out of the arena of quantum physics. Actually, they did point out the violations of total probability law that can be noticed in experimental findings in biology, which characteristics have also been observed while studying several contextual phenomena in the cognitive domain. In explaining these aspects, i.e., a situation or "context", they started with an example from modern physics, i.e., the case of the double-slit experiment, but with both slits open. This context is not a simple sum of the two contexts, say, S_i for $i = 1$, 2 and $S_{1 \cup 2}$ for both the slits open, i.e.:

$$S_{1 \cup 2} \neq S_1 \cup S_2$$

In quantum mechanics, the violation of the law of classical probability arises from the difference between the structures of two contexts: $S_{1 \cup 2}$ and $S_1 \cup S_2$.

Asano et al. (2013) took an example from the domain of cognitive psychology, which is as follows: Chocolates are given to the subjects who are asked whether chocolates are sweet ($C = 1$) or not ($C = 2$). Then, the statistical data obtained are used to determine the probabilities $P(C = 1)$ and $P(C = 2)$. Next, another set of experiments is done, i.e., at first sugar was given to the subjects before distributing the chocolates. Collecting the statistical data, the probabilities $P(S = 1)$, $P(S = 2)$, $P(C = 1|S = 1)$ and $P(C = 1|S = 2)$ are estimated. When estimated, the probability that the chocolates is sweet is given by:

$$P(C = 1|S = 1)\ P(S = 1) + P(C = 1|S = 2)\ P(S = 2)$$

Now, whereas, it is easy to show that this is equal to $P(C = 1)$, using the rule of classical probability theory it can easily be shown that the above probability is smaller than $P(C = 1)$.

$$P(C = 1)\ P(C = 1|S = 1)\ P(S = 1) + P(C = 1|S = 2)\ P(S = 2)$$

The left hand side probability $P(C = 1)$ is estimated in the context *"subjects did not taste sugar before taking chocolates $P_{S-sug}(C = 1)$. Contexts: 'a tongue tasted sugar (P_{Ssug})' and 'a tongue did not taste sugar $P_{S-sug}(C = 1)$' are different."*

Intuitively, the result of $P_{S-sug}(C = 1) \neq P_{Ssug}(C = 1)$ seems to be quite easy and obvious, but to explain these results mathematically, a properly developed probability space is needed. It is interesting to note that Asano et al. reached a very crucial and interesting conclusion from their experimental observation, i.e., enormous knowledge about the physical and chemical structure of the tongue is essential to develop a proper probability space.

So, essentially, further works are needed to shed new light on the applicability of this type of context dependence on experiments in cognitive neuroscience. In the next chapter, the quantum logic, in the context of neuroscience, will be discussed from a heuristic point of view.

References

Aerts, D. and Steirteghem, B. Von (1999) Int. Journ. Theor. Phys.; **39**(3), 497-502.

Asano, Masanari et al (2013) Found. Phys. 43(7);895-911.

Birkhopf, G. & von Neumann, J. (1936) Annals of Mathematics, **37**, pp 823-843.

Boole George, Frier (1854) *An Investigation of Laws of Thought: on which are founded the Mathematical theories of Logic and Probabilities*; Google.co.in.

Constantin, Piron. (1976) *Foundations of Quantum Physics, Reading MA*; W.A. Benjamin Inc., Massachusettes.

Cushing, J.T. (1967) American Journal of Physics; *35: 858–862.*

Giulini, Domenico (2001); *"Das Problem der Tragheit (PDF)"*; Preprint; Max-Planck InstitutfürWissenschaftsgeschichte; **190**: 11–12, 25–26.

Griffiths R.B. (2009) Consistent Histories. In Daniel Greenberger; Klaus Hentschel, and Friedel Weinert, editors, Compendium of Quantum Physics, pages 117–122. Springer-Verlag, Berlin, 2009.

Griffiths, R.B. (2011) *A Consistent Quantum Ontology;* arXiv:1105.3932v2

Heelan, Patrick, A. (1970); Foundation of Physics, **1** (2):95-110.

http://users.mat.unimi.it/users/galgani/arch/heis25ajp.pdf;

Kenny, A. (1976) *Thomas D"Aquinas- Logic and Metaphysics;* University of Notre Dame Press.

Lange, Ludwig (1885) *Ueber die wissenschaftilicheFassung des GalilichenBeharrungsgesetzes* ; Philosophische Studien; **2**: 266–297.

Mach, Ernst (1883/1912) *Die Mechanik in ihrerEntwicklung (PDF)*; Leipzig: Brockhaus,(first English translation in 1893).

Mittlesteadt, Peter (2011); *Rational Reconstructions of Modern Physics*; Springer.

Poincaré, Henri (1889) *Théoriemathématique de la lumière;* **1;** Paris: G. Carré& C. Naud Preface
    partly reprinted in "Science and Hypothesis", Ch. 12; (1901a); *"Sur les principes de la
    mécanique"*, Bibliothèque du Congrès international de philosophie: 457–494.
Quinne, W. V.O. (1951) The philosophical Review; **60**;1(Jan 1951), 20-43; (1976); *"TwoDogmas
    of Empiricism: Can Theories be Refuted"*? in *"The ways of paradox and other essays"*;
    books.google.com

Chapter 9
Modern Neuroscience and Quantum Logic

I believe that at the end of the century the use of words and general educated opinion will have altered so much that one will be able to speak of machines thinking without expecting to be contradicted.
—Alan Turing, Computing machinery and intelligence
(Computing machinery and intelligence. Mind, 59, p. 7)

Abstract The decision making from a neuroscience perspective is one of the outstanding problems in twenty first-century science. The evolution of higher brain function has given us the capacity for flexible decision making. Gerstner et al. (Neuronal Dynamics: From Single Neurons to Networks and Models of Cognition. Cambridge University Press, Cambridge, 2014) made an attempt to understand decision making based on an interacting neuronal model. According to their proposal, various neuronal populations with different options take part in a competition, and the population with the highest activity wins that competition. Here, the group of strongly connected neurons is considered to play the important role in making the choice. This procedure of Gerstner has been elaborated in this chapter to understand decision making. Next, we address the neuronal architecture necessary for implementing the logic called *"quantum logic"*. In fact, it becomes necessary and important to find out which neuronal circuitry in this architecture is responsible for decisions and, at the same time, what are the underlying processes they follow in arriving at a decision. McCollum (Systems of Logical systems; Neuroscience and quantum logic: Foundation of science, Springer, Berlin, 2002) made an attempt to understand and apply quantum logic within the framework of modern neuroscience. It is noted that he is more realistic in his approach to understanding decision making and brain function. He made the interesting observation in that, in the beginning, one does not need to assume all kinds of mathematical formalism like Hilbert space structure, quantum probability, etc. Instead, it is worthwhile to study the functioning of the neuronal architecture first, and then, as the next step, to look for a suitable logic or mathematical tools to explain the observed results. But the central issue, i.e., *"Where is the decision taken"*, still remains a mystery.

Keywords Decision making · Quantum logic · Interacting neuron · Stochastic equation

© Springer India 2016
153
S. Roy, *Decision Making and Modelling in Cognitive Science*,
DOI 10.1007/978-81-322-3622-1_9

The evolution of higher brain function has given us the capacity for flexible decision making. The evidence we obtain through our senses (or from memory) need not precipitate an immediate, reflexive response. Instead, our decisions are observed to be deliberately dependent and provisional, contingent on other sources of information, long-term goals and values. Recently, Gerstner et al. (2014) described the decision-making process using a network of interacting neurons. According to their proposal, various neuronal populations with different options take part in a competition and the population with the highest activity wins the competition. Here, the group of strongly connected neurons plays an important role in making the choice. We have already indicated some results related to visual perception that are hard to explain using classical probability theory. From a neuroscience perspective, Gerstner et al. (2014) studied many perceptual phenomena as part of decision making. Their results suggested that the capacity for higher brain function might be dependent on the brain's ability to accumulate and then to combine it with the preservation of those information over time. But, as presented earlier, we stumble on the central issue, i.e., the answer to *where is the decision taken*" is still a mystery. According to their hypothesis, such persistent activity is nothing but the process of integration with respect to time, which happens due to the process of accumulation and storage of the associated information. They based this on an analogy to the findings of Robinson and Fukusima et al. (Robinson 1989; Fukusima et al. 1992), i.e., on the neural integration in the brain stems that converts eye velocity to position. According to their suggestions, the cortical neural integrators make use of the combination of sensory data. Correspondingly, these integrators generate an evolving conception about the state of the world which, then, ultimately is used in the planning of appropriate behavior. In their experiments, taking rhesus monkeys as subjects, they explored extensively this idea about the accumulation of sensory information with respect to time that underlies a simple perceptual decision.

Roitman and Shadlen (2002) made an interesting observation about the lateral intra-parietal (LIP) area; they studied the effects of moving random dot stimuli in the case of perceptual decision making. The LIP area is located in the visual-processing stream. Their experiment can be explained using a simple decision-making model where various neuronal populations compete with each other through shared inhibition. In any case, we are not going to discuss this experiment here in detail. Though it has been already been dealt in detail by Gerstner et al., as mentioned above, a brief outline of their work is offered below. They considered a network of two excitatory populations of spiking neurons, interacting with inhibitory neurons. Here, the excitatory neurons are connected randomly with weight w_{EE} and that of inhibitory neurons with weights w_{IE} and w_{EI}, respectively, i.e., for connections to and from inhibitory neuronal populations. These weights and neuronal parameters are adjusted in such a way that all neurons exhibit spontaneous activity in the absence of external input. A mathematical model of three populations is considered as follows:

Let the population activity of an excitatory population $A_{E,k}$ be $A_{E,k} = g_E(h_{E,k})$ where as that of the inhibitory potential is $A_{inh} = g_{inh}(h_{inh})$. Let the gain functions of excitatory and inhibitory neurons be denoted by g_E and g_{inh}, respectively. Then, the evolution of input potential can be written as:

$$\tau_1 \frac{dh_{E,1}}{dt} = -h_{E,1} + w_{EE}g_E(h_{E,1}) + w_{E1}g_{inh}(h_{inh}) + Rl_1$$

$$\tau_2 \frac{dh_{E,2}}{dt} = -h_{E,2} + w_{EE}g_E(h_{E,2})w_{E1}g_{inh}(h_{inh}) + Rl_2$$

$$\tau_{inh} \frac{dh_{inh}}{dt} = -h_{inh} + w_{IE}g_E(h_{I,1})w_{IE}g_E(h_{E,2})$$

These equations are very complicated to solve. With some assumptions, they can be reduced to two-dimensional equations, and then it becomes easy to solve these two-dimensional equations in phase-plane. In doing so, the following two assumptions are made:

1. The membrane time constant of inhibition is shorter than that of excitation, $\tau_{inh} \ll \tau_E$.
2. The gain functions of inhibitory neurons are considered as linear, i.e., they can be written as $G_{inh}(h_{inh}) = \gamma \cdot h_{inh}$, taking $\gamma > 0$.

Applying the above two assumptions, we obtain another two differential equations:

$$\tau_E \frac{dh_{E,1}}{dt} = -h_{E,1} + (w_{EE} - \alpha)g_E(h_{E,1}) - \alpha g_E(h_{E,2}) + Rl_1$$

$$\tau_E \frac{dh_2}{dt} = -h_{E,2} + (w_{EE} - \alpha)g_E(h_{E,2}) - \alpha g_E(h_{E,1}) + Rl_2$$

This procedure replaces a model of three populations by two excitatory populations interacting via an effective coupling strength, given by:

$$\alpha = -\gamma w_{IE}w_{EI} > 0.$$

The derivation clearly indicates the existence of mathematically equivalent descriptions where explicit inhibition by inhibitory neurons is replaced by effective inhibition between excitatory neurons. This concept of effective coupling enables us to think of competition between neuronal groups in a more comprehensive manner. It helped the above mentioned authors to make it possibly to construct a framework for dynamic decision making.

Within this framework of phase plane analysis, there exist other models of decision making, too. One of them is based on the gradient descent method, where:

$$\frac{dx}{dt} = -\eta \frac{dE}{dx}$$

with positive constant η.

In such a case, the dynamics of the decision process can be formulated as gradient descent. During a short time interval, the decision variable moves and the movement turns to the left for positive slope. Hence the movement is always downhill and the energy decreases along the trajectory. The change of energy can be calculated along this trajectory. Here, the energy plays the role of Lyapunov function—the quantity which does not increase along the trajectory of a dynamical system. The change of energy along the trajectory can be written as

$$\frac{dE(x,t)}{dt} = \frac{dE}{dx}\frac{dx}{dt} = -\eta \left(\frac{dE}{dt}\right)^2 \leq 0$$

This depiction of the energy framework is analyzed in phase space with two-dimensional models. Apart from the gradient descent model, the drift-diffusion model has also been studied. The drift-diffusion model (Ratclif and Rouder 1998) is a basically phenomenological model to describe the preference of choices and the distribution of reaction time a binary mode of decision-making tasks. However, decision making is ultimately reduced to understanding the cellular basis of cognition. No such theory has so far been developed from a neuroscience perspective. In these circumstances, it is possible to discuss the neuronal architecture necessary for implementing the logic called "*quantum logic*". Of course, then arises the necessity of finding out which neuronal circuitry in this architecture is responsible for decision making. Future research may help us to clarify the present situation.

McCollum (2002) made an attempt to understand quantum logic within the framework of modern neuroscience. As the neuronal architecture is organized in many hierarchical levels, his study pointed out that:

> Intrinsic dynamics of the central vestibular system (CVS) appear to be at least partly determined by the symmetries of its connections. The CVS contributes to whole-body functions such as upright balance and maintenance of gaze direction. These functions coordinate disparate senses (visual, inertial, somato-sensory, auditory) and body movements (leg, trunk, head/neck, eye). They are also unified by geometric conditions. Symmetry groups have been found to structure experimentally-recorded pathways of the central vestibular system. When related to geometric conditions in three-dimensional physical space, these symmetry groups make sense as a logical foundation for sensorimotor coordination. (McCollum 2002)

This indicates that the intricate organization level invites certain kind of mathematics that reflect its logical structure. The developments of mathematics may help us to choose a particular type of mathematical tools depending on the organizational structure. This is similar to the use of non-Euclidean geometry to understand gravitation by Einstein. It is to be noted that, though non-Euclidean geometry was developed by mathematicians long before Einstein's work, it was he who first applied this mathematical framework to understand the characteristics of gravitation. It is worth mentioning, there exists two approaches regarding the relationship between scientific disciplines like physics and biology. If biology is considered to be a derived branch of physics, then we do not need separate foundational concepts and mathematical structures for biology other than those inherited from physics. On the other hand, if biology is considered as separate scientific branch from than physics but interacting with it, then the logical systems developed for biology are to be modelled directly so as to establish its own foundational concepts and mathematical foundations. McCollum studied the various characteristics of the neuronal organizations and the necessity of suitable mathematics, instead of applying the mathematical tools directly in the context of quantum logic.

Bialek (1987) discussed the applicability of quantum mechanics at the synaptic level for contributing stochastic properties in all chemical systems. He raised a question about the limit of the sensory systems what is needed for reaching the physical limits to fulfill their performance. So far as applicability is concerned, there are noted facts which have indicated that the usually considered broad classes of various, even fruitful mechanisms, simply cannot reach these limits. They noted that, in favorable cases, within the receptor cell, very specific requirements are placed on the mechanisms of filtering, transduction and amplification. These ideas helped to prove the relevancy of the quantum theory of measurement in understanding the physical limits to sensation and perception range. They also made some observational comments in their studies about the physical limits to sensory performance. The results did change their views of the sensory systems, and they suggested that both the transduction mechanisms in receptor cells and that of the neural mechanisms are responsible for processing sensory information.

McCollum (2002) made detail analyses of the applicability of quantum mechanics and the use of multi-valued logic at the level of neuronal mapping and sensory motor coordination. Recently, Kramer (2014) posited a contrarian view and proposed a semantic of the standard quantum logic (QL) based on Boolean algebras. He suggested classical logic as a completion of the quantum logic QL. In other words, Birkhoff and von Neumann's classic thesis has been criticized for depicting the logic (the formal character) of quantum mechanics to be non-classical in nature, the same as happened to Putnam's thesis (Putnam 1975) for considering that quantum logic (of his kind) would be the correct logic for propositional inference in general. The propositional logic of quantum mechanics is such a modal but classical one, and the correct logic for propositional inference need not possess an extroverted quantum character. One normal necessity for the modality suffices to capture the subjectivity of observation in quantum experiments. The key to their results is based on the translation of quantum negation as classical negation of observability.

The investigations are still in their infancy, and it is not at all clear whether quantum logic, the propositional calculus and Hilbert space formalism can be realized in understanding neuronal architecture. Future studies may shed light and further enrich our understanding of the issues connected to the context dependence, as well as to the abstract, framework of quantum logic in neuroscience.

References

Bialek, W. (1987) Annual Review of Biophysics and Biophysical Chemistry; **16**:455-478.

Fukusima, K., Kaneko, C & Fuchs, A. (1992) Prog. Neurobiol.; **39**:609-639.

Gerstner, Wulfram et al (2014) *Neuronal Dynamics: From Single Neurons to Networks and Models of Cognition*; Cambridge University Press.

Kramer, Simon (2014); A Galois-Connection between Myers-Briggs' Type Indicators and Szondi's Personality Profiles arXiv: 1403. 2000 v1[cs.CE]

McCollum, Gin (2002) *Systems of Logical systems*; *Neuroscience and quantum logic*: *Foundation of science*, Springer;

Ratclif R. & Rouder J. (1998) Psychol. Sci.; **9**; 347-356.

Robinson, D.A. (1989) Ann. Rev. Neurosci.; **12**; 33-45.

Roitman J. and Shadlen M. (2002) J. Neurosci., **22**, 9475-9489.

Putnam, H. (1975) *What is mathematical truth? ; Historia Mathematica"*; Elsevier.

Chapter 10
Future Directions of Modelling the Uncertainty in the Cognitive Domain

There is a real world independent of our senses; the laws of nature were not invented by man, but forced on him by the natural world. They are the expression of a natural world order
—Max Planck (1936)

Abstract Recent experimental findings clearly suggest that classical probability theory is still not successful to explain modalities in human cognition, especially in connection to decision making. The major problem seems to be the presence of epistemic uncertainty and its effects on cognition at any time point. Moreover, the stochasticity in the model arises due to the unknown path or trajectory (the definite state of mind at each point in time) a person follows. In fact, there exists more ambiguity than clarity in some aspects of cognition, for example, regarding cognitive affective decision-making behavior and subsequent choices (George et al. in Academy of Management Review 31: 347–365, 2006). The concept of ambiguity, its typical role and its function are not only long-standing debatable concerns in the humanities but also assume a similar level of importance, and are the subject of much debate, in contemporary neurological and cognitive approaches. For example, biologist Fredrick Grinnell made an argument clarifying that ambiguity "is inherent in carrying out and reporting research" (Grinnell in Science and Engineering Ethics 5:205, 1999). Not only that, but he also pointed out some parameters of ambiguity that support his view, for instance, the difficulty of distinguishing between data and noise in research "at the edge of discovery" (ibid. 207). Interestingly, as a positive side of ambiguity, he observed "while ambiguity in method and procedure—as an aspect of data and its interpretation—is clearly a potential problem, posing the risk of distortion, it is also associated here with creative insight." The consideration of *"black box model of human mental functions"* produces much ambiguity. So, as we are aware of the intertwined mathematical, neurological and cognitive mysteries of the brain, ultimately, these very characteristics are challenges faced by the human cognition and give rise to a very complicated and complex integrative system of cognitive computation and affective perception. A generalized version of probability theory, borrowing the idea from the quantum paradigm, may be a plausible approach. Quantum theory enables a person to be in an indefinite state (superposition state) in the context of neurobiology, especially in relation to central nervous

© Springer India 2016

S. Roy, *Decision Making and Modelling in Cognitive Science*,

DOI 10.1007/978-81-322-3622-1_10

system (CNS), and this allows all these states to be potentially (of course, with proper probability amplitudes) expressible at each moment. Thus, a superposition state seems to provide a better representation of the conflict, ambiguity or uncertainty that a person experiences at each moment. This situation can be described in somewhat poetic language, i.e., the dynamical laws, as demonstrated by nature, allow equations but are only sometimes posed by the Mind of the experimentalist. This way, consciousness (as expressed in his introduction to consciousness or cognition can be expressed as only a critical description of human activity and not as a theory of mind. Some critical remarks about handling the ambiguity and contradictions due to emotions within quantum probability framework are made in this chapter. Mental states and related decision making have been widely discussed in various ancient and medieval Indian traditions. The existence of an indefinite state of mind was discussed by many Indian philosophical schools, including Buddhism. The concept of the neutral mind, as well as equanimity, is discussed in this chapter, especially because it has a potential to be related to modern neuroscience, as well as to quantum theory.

Keywords Affective computing · Quantum probability · Indefinite state of mind · Neutral mind · Equanimity · Buddhist school · Samatvam

10.1 Remarks on Affective Computing and Quantum Probability

The concept of affective computing is a relatively recent addition to the vocabulary of the computing world. Broadly speaking, it describes computing that is somehow connected to emotion—sometimes known as *emotional artificial intelligence.* Logic is usually embodied in a device like a computer. But, as we all are aware, feelings of emotion are essentially about how emotional action programs are perceived. The most challenging part is how a computer recognizes emotion and what the embedded logic should be which will successfully exploit both artificial information and that from emotions simultaneously so that we would achieve a successful outcome. Dr. Rosalind W Picard first coined the term *affective computing in her book "Affective Computing" in 1997.* Her motivation was to create a machine for which the programs should be written in such a way that 'affect' is included, so as to enable interactions with people. In this novel framework, the emotional interactions between human beings and machines are described in terms of embedded logic. Some attempts have been made by Damasio (2010), who highlighted pointed three points connected to dealing with this problem, i.e.:

1. Triggering system
2. Neural system
3. The actions

that all together constitutes the emotion.

Etkins et al. (2015) did put forward an interesting proposal understanding the neural bases of diverse forms of emotion regulation. They performed substantial neurological studies which identified several brain regions that play a part in the regulation of emotions. They found that these include the cingular, ventromedial prefrontal cortices. Moreover, they also reported on the substantial roles played by lateral prefrontal and parietal cortices at the same time. Based on their findings, they proposed a unifying conceptual framework that enables us to have an idea about the neural bases of diverse forms, which are responsible for regulation of emotions. Since emotions shape our thoughts, behaviors and feelings, they must be regulated. So, in order to implement this idea in the context of man-machine understanding, as a first step, one needs to model the emotions and then the associated logic of the mental process. In the next step, it is necessary to embody this logic in the machines, so as to understand man-machine interaction.

In this book, it has been emphasized that the use of quantum logic (a kind of paraconsistent Logic) (Priest 2002) is the most relevant one to model emotion and its various contradictory aspects. Normally, contradictory aspects related to emotion cannot be accommodated within the realm of the Aristotelian framework (developed by Boole, hence Boolean logic). But the generalized logical framework (known as paraconsistent logic) does not eliminate the contradictions but incorporates them in a logical way. As a definition, this logic can be stated as paraconsistent '*if*' its relations related to consequences are not explosive, or stated another way, if it denies the contemporary logical orthodoxy. It means that, if a relation has logical consequences, defined either semantically or proof-theoretically, it definitely demonstrates paraconsistency. Together with these conditions, even under the circumstances that available information is inconsistent, the inference relation does not explode into a '*triviality*'. Thus the positive aspect of this logic lies in the fact that it accommodates inconsistency in a sensible manner, i.e., it treats inconsistent information as considerable information that can be expressed as a property of a consequence relation, $\{A, A\} \vDash B$ for every A and B (*ex contradictione quodlibet* (ECQ)). Quantum logic as described in this book can be described to be a type of paraconsistent logic. It then becomes of primary importance to consider and analyze first the details of mental functions. Then only does the corresponding modelling becomes necessary to understand the man–machine interface and affective computing from a more realistic perspective.

10.2 Epistemological Issues

As mentioned in preceding chapters, the recent experimental findings have not yet clearly decided whether classical probability theory is capable to model successfully human cognition, especially in connection to decision making. The major problem seems to be the presence of epistemic uncertainty and its effects on cognition at any time point. Moreover, the stochasticity in the model arises due to the unknown pathway or trajectory (the definite state of mind at each point in time) a person follows. A generalized version of probability theory, borrowing the idea from the quantum paradigm, may be a plausible approach, because quantum theory allows a person to be in an indefinite state (superposed state) within the context of neurobiology, especially in the context of the central nervous system (CNS) (McCollum 2002). Not only that, but this theory also enables all these states to be potentially (with an appropriate probability amplitude) expressed at each moment (Heisenberg 1958). Hence, a superposition state seems to provide a better representation of the conflict, ambiguity or uncertainty that a person experiences at each moment (Busemeyer et al. 2011). Conte et al. (2009a, b) performed a set of very important experiments that demonstrated that mental states follow quantum mechanical features during the perception and cognition of ambiguous figures.

However, the framework of quantum probability considering the superposition of mental states is an abstract framework devoid of material content like the concept of elementary particles and various fundamental constants in nature, for example, the Planck constant (h), the speed of light (c), and the gravitational constant (G) in modern physics. This framework can be applied to any branch of science dealing with decision making, such as biology and the social sciences. Very few attempts have been made so far in the context of neuroscience and higher order cognitive activities. We have already discussed quantum logic in the context of neuroscience in previous chapter.

For centuries, the various mental states and decision making have been widely discussed in various Indian traditions. At first we will briefly discuss various states of minds from the Buddhist perspectives, as present in *Abhidamma* (Kāśyapa (ed.) 1982) studies, the feelings are analyzed mainly into two categories—by intrinsic nature and by the governing faculty.

By the way of intrinsic nature, the **feeling** can be divided into three categories:

1. Pleasant feeling—'*sukha*'
2. Painful feeling—'*dukkha*'
3. Feeling that is neither painful nor pleasant—'*adukkham-asukha*'

The **feeling**, when analyzed from the angle of the governing faculty, is divided into five categories:

1. Pleasure—'*sukha*'
2. Pain—'*dukkha*'
3. Joy—'*somanassa*'
4. Displeasure—'*domanassa*'
5. Neutral feeling—'*upekkhā*'

Again, the **citta or mind** can be classified into the following categories depending on the three categories of feeling as described in the context of the intrinsic nature of mind. They are:

1. *Citta* with pleasure—'*sukha-sahagata citta*'
2. *Citta* with pain—'*dukkha-sahagata*'
3. *Citta* with neither—'pleasure-nor pain'

Depending on the above five categories of feeling associated with the governing faculty, **citta** or **mind** can again be classified as:

1. *Citta* with pleasure—'*sukha-sahagata*'
2. *Citta* with pain—'*dukkha-sahagata*'
3. *Citta* with joy—'*somanassa-sahagata*'
4. *Citta* with displeasure—'*domanassa-sahagata*'
5. *Citta* with neutral feeling—'*upekkhā-sahagata*'

With the help of above system, now we are able to discuss decision making and judgment with the help of concepts such as states of mind like *Citta* with 'neither-pleasure-nor pain' and *Citta* 'with neutral feeling' as *upekkhā-sahagata*. The similarity of these states can be drawn with the intermediate states associated with the superposition of complementary states in quantum theory. It is important to highlight the scientific aspects of these states, which are conducive to proper decision making by an individual.

The challenging issue in modern neuroscience is how the neurons in the brain operate, so as to interpret the above kinds of intermediate states that give rise to the superposition of two complementary aspects. For example, it is possible to understand the operation "yes" or "no" with the operations performed by neurons. But, if there exists an interference term due to superposed states of mind, then one can think of an intermediate term corresponding to a state like:

$$a \, |\text{HAPPINESS}\rangle + b \, |\text{UNHAPPINESS}\rangle$$

where a and b are positive constants. Here $|\text{HAPPINESS}\rangle$ and $|\text{UNHAPPINESS}\rangle$ are two vectors representing the happy and unhappy states of mind, respectively. According to the superposition principle, one can think of a state as neither happy nor unhappy. How the brain interprets this operation on the cellular level (i.e., at the level of neuron) is hard to conceive. In quantum theory, such superposed states have been discussed and have raised much interest in the scientific community. The famous physicist Schrödinger (1935) formulated this problem as the "Cat Paradox," that is, one can think of a superposed state of the 'living state' and the 'dead state' of a cat, but only for an observer will it be either in the living or dead state. The observation procedure is considered as a classical process, and hence the superposed state or the wave function collapses to either of the states. Shimony (1993) proposed a "potentiality interpretation," i.e., the superposed state function has the potential to be in either of the complementary states.

The type of issues, mentioned above, has been also debated in various other schools of Eastern philosophy. This kind of intermediate states of mind can be analyzed in relation to the state of mind known as "**equanimity**" in Eastern philosophy. The ancient Sanskrit word "**upeksha**" or "*upekkhā*" in Pali is used similarly to the word equanimity, as one of the four sublime states of mind in Buddhist philosophy or sublime attitude in Patanjali's '**Yoga Sutras**' (Patanjali et al. 1914 (translation)).

The term equanimity comes from Latin "*aequanimitas*". The word "*aequanimitas*" originates from "*aequus*" (equal) and "*animus*" (mind) which mean "*calmness and composure, especially in a difficult situation*" (*Oxford English Dictionary*). In fact, most of the spiritual traditions of the world consider equanimity as central to their teachings.

In Bhagavad Gita, the concept of equanimity is described as yoga in the following way: "*Perform action, O Arjuna, being steadfast in yoga, abandoning attachment and balanced in success or failure. Such equanimity is called yoga.*" (Samatvam: *The Yoga of Equanimity, Yoga Publication Trust, Munger, 2009, p. 5*).

Sivananda (2000) analyzed the situation in detail and defined **Samatvam** as the equanimity of mind and outlook and equipoise. It is the state where the human being will be able to keep the mind steady and balanced in all possible conditions of life and as well, will be able to think and make a decision, being in an absolute tranquil and equipoised state of mind.

In Buddhist tradition, the term "equanimity" has been defined in various ways—the core of this definition rests on "gazing upon" or "observing without interference." In the *Theravadan Buddhist School*, two main definitions of the term "equanimity" are used. The first one refers to "neutral feeling"—a mental experience that is neither happy nor unhappy. This first usage of equanimity corresponds to the Western psychological notion of "neutral valence" and is commonly experienced throughout any ordinary day. In this state, mind is considered to be capable of reaching a decision and making a judgment. The second usage of the term "equanimity" refers to a state of mind that cannot be "swayed by biases and preferences" (Anuruddha and Bodhi 2000).

Now, similar to the potentiality interpretation of quantum theory, we can think of such a state of mind where the mind is capable of decision making even while remaining in a state, superposed of two complementary aspects. As explained earlier, the process of decision making is a measurement process that occurs at certain hierarchical level of the neuronal architecture present in central nervous system. So, even when this state of mind can be considered as a superposed state of complementary aspects, potentiality always remains for the actualization of one of the complementary states. Then, it raises the possibility of describing the state of mind called "equanimity" using the framework of quantum logic where a decision or judgment can be made without any bias or preference. It is worth mentioning that Buddhist, as well as Hindu, philosophers agree on the issue that one needs substantial practice to achieve the state of mind called "equanimity." It is now clear from the above analysis that the abstract framework of quantum logic in the context of neuroscience may help us to understand the various states of mind like 'equanimity', in which decision making happens with unbiased judgment. It opens up a new dialogue between modern science and ancient Indian philosophy, as the latter is likely to offer an explanation to the superposed state, which phenomena have been under intensive, ongoing study, by the scientific community in quantum theory.

References

Anuruddha, B. & Bodhi, B. (2000) *A comprehensive manual of Abhidhamma: the Abhidhammattha sangaha of Ācariya Anuruddha* (1st BPS Pariyatti ed.); Seattle.

Busemeyer, J.R., Pothos, E.M., Franco, R., & Trueblood, J. S. (2011); Psychol. Rev., **118** (2), 193-218

Conte, Elio et. al. (2009a); Open system and Information dynamics, **16**, 85.

Conte E., Khrennikov, Y., Toderello, O. et al.; (2009b); Open system & Information Dynamics, **16** (1),1-17.

Damasio Antonio (2010); Neural basis of emotions; Scholarpedia; **6**(3);1804.

Etkins Amit, Buchel C. & Gross, J.J. (2015); Nature Reviews Neuroscience, 16, 693-700 (published on line 20 October 2015).

George E., Chattopadhyay P., Barden J. (2006) Academy of Management Review **31**, 347-365.

Grinnell, F. (1999) Science and Engineering Ethics, **5**: 205–214.

Heisenberg (1958) *Physics and Philosophy*; New York, Harper.

Kāśyapa Jagadīśa Bhikku (1982) (ed.) *The Abhidhamma Philosophy, or, the Psycho-Ethical Philosophy of Early Buddhism*; Bharatiya Vidya Prakashan.

Max Planck (1936) *Philosophy of physics*, W.W. Norton & Company, inc., p. 46

McCollum, Gin (2002) *Systems of Logical systems*; "*Neuroscience and quantum logic*: Foundation of science*, Springer.

Patanjali, Vacaspati Misra, Vyasa, and James Haughton Woods (1914); *The Yoga-System of—or the ancient Hindu Doctrine of concentration of mind embracing the mnemonic rules, called Yoga-Sutras, of—and the comment, called Yoga-Bhashya (etc.);* Harvard University Press.

Priest, G. (2002) *Paraconsistent Logic; Handbook of Philosophical Logic (2nd edition);*vol 6; gabby & F. Guenthner (eds); Dordrecht: Kluwer Academic Publishers; pp 287-393.

Schrödinger, Erwin (1935) *Die gegenwärtige Situation in der Quantenmechanik (The present situation in quantum mechanics).* Naturwissenschaften; **23** (49): 807–812.

Shimony Abner (1993); "*The Family of Propensity Interpretations of Probability*," Lecture at University of Bologna, 1988 Box 1, Folder 19 Abner Shimony Papers, 1947-2009, ASP.2009.02, Archives of Scientific Philosophy, Special Collections Department, University of Pittsburgh.